CliffsQuickReview® Earth Science

By Scott Ryan

Wiley Publishing, Inc.

About the Author

Scott Ryan teaches Earth Science at Ardsley Middle School, a National Blue Ribbon School of Excellence, in New York.

Author's Acknowledgments

Greg Tubach: For finding me and his support

My family: For their love and understanding

Kelly, Lisa, Lori, and Suzanne: For their excellent feedback and corrections

Sabrina, Rich, Jay Levine, Al Gore: For the Internet

Publisher's Acknowledgments

Editorial

Acquisitions Editor: Greg Tubach

Project Editor: Kelly D. Henthorne

Copy Editor: Lori Cates Hand

Technical Editor: Lisa Berkowitz

Production

Indexer: Anne Leach

Proofreader: Jacqui Brownstein

Wiley Publishing, Inc. Composition Services

CliffsQuickReview® Earth Science
Published by:
Wiley Publishing, Inc.
111 River Street
Hoboken, NJ 07030-5774
www.wiley.com

Published by Wiley, Hoboken, NJ
Published simultaneously in Canada

Library of Congress Cataloging-in-Publication Data

Ryan, Scott.
Cliffsquickreview earth science / by Scott Ryan.
p. cm. -- (Cliffs quick review)
Includes index.
ISBN-13: 978-0-471-78937-6 (pbk.)
ISBN-10: 0-471-78937-2 (pbk.)
1. Earth Science--Outlines, syllabi, etc. I. Title. II. Title: Earth Science. III. Series.
QE41.R93 2006
550--dc22

2006008778

ISBN-13 978-0-471-78937-6
ISBN-10 0-471-78937-2
Printed in the United States of America
10 9 8 7 6 5 4 3
1O/SU/QU/QW/IN

Table of Contents

Chapter 1

INTRODUCTION TO EARTH SCIENCE

Chapter Checkin

- ❑ Knowing the basic ways to make a measurement
- ❑ Understanding the different ways to graph data
- ❑ Calculating the percent error of a measurement

The Earth is the only planet in our solar system that has an atmosphere that can support life. Water can be found in all three states of matter (liquid, solid, and gas) and Earth also has a dynamic crust. Forces are constantly at work changing the landscape on the Earth's surface. Earth Science can be divided into four branches. Geology is the study of rocks, minerals, and forces that wear down the surface and build mountains. Meteorology covers weather, climate, and the atmosphere. Astronomy investigates planets, stars, and other features outside of the atmosphere. The study of the oceans is called oceanography.

Observations and Measurement Methods

An **observation** is made when you collect data through your senses. **Instruments** such as telescopes, microscopes, meter sticks, and stopwatches are used to extend your senses. An **inference** is an educated guess based on data that is collected and interpreted.

Basic Units of Measurement

You can make measurements in five basic ways. A measurement consists of a number and a unit. The five basic measurements, units, and instruments used to make these measurements are in Table 1-1.

Table 1-1 Units of Measurement

Measurement	*Basic Unit*	*Instrument Used*
Length	Meter	Meter stick
Mass	Gram	Triple beam balance, electronic balance
Volume	Liter	Graduated cylinder, overflow canister
Time	Second	Stopwatch
Temperature	Degrees Celsius	Thermometer

Derived Units

From these basic units, **derived units** can be formed. Density is an example of a derived unit that is used in Earth Science. Density = mass/volume and the units are grams/cm^3 or grams/ml.

Scientific Notation

Scientific notation is used to show very small or very large numbers in a convenient way. This consists of a base number that can range from 1.0 to $9.\overline{999}$. This base number is multiplied by 10^x. If the value of x is negative, then the original number is less than 1.0, and the decimal place is moved to the left. A positive number in the exponent place means that a large number is converted to scientific notation, and the decimal place is moved to the right. Examples of the use of scientific notation are as follows:

43,000,000,000,000,000 mi becomes 4.3×10^{16} mi.

.00152 cm becomes 1.52×10^{-3} cm.

Prefixes can be added to the base units to subdivide a measurement. Table 1-2 shows prefixes, their abbreviations, and their multiplying factors.

Table 1-2 Scientific Notation Prefixes, Symbols, Multiplying Factors

deci	d	1/10	deka	da	10
centi	c	1/100	hecto	h	100
milli	m	1/1,000	kilo	k	1,000
micro	μ	1/1,000,000	mega	M	1,000,000
nano	n	1/1,000,000,000	giga	G	1,000,000,000

Percent Error

You can determine the inaccuracy of a measurement when compared to the actual measurement by finding the **percent error.** This is also called the percent deviation. This tells you how far away your value is from the accepted value, in a percent form. The formula used is

$$\text{Percent error} = \frac{\text{actual value} - \text{your value}}{\text{actual value}} \times 100$$

Density

The **density** of an object is its mass divided by its volume. Mass is the amount of matter in an object. Volume is the amount of space that an object takes up. Volume can be measured in cm^3 for solids and in milliliters for liquids. One cm^3 equals 1 ml. This value is a physical characteristic for a substance that does not change if the size of the material changes. The reason for this is shown in the following paragraph:

$$D = m/v$$

If the mass is 12 g and the volume is 6 cm^3, the density is 2 g/cm^3. Suppose that the object was cut in half. The mass would be 6 g and the volume would be 3 cm^3. By using the density formula again, the resulting density is the same, 2 g/cm^3. The formula is just a ratio of matter ("stuff") to volume ("space"). If you change both equally, the density will remain the same. If, in a different example, the volume is changed and the mass remains the same, then the density changes. See Figure 1-1 for this concept in more mathematical detail.

Figure 1-1 Changes in density shown mathematically.

$$D = \frac{M}{V} \quad D = \frac{12g}{2cm^3} = 6\ g/cm^3$$

If volume decreases → $V = 1cm^3$

$$D = \frac{M}{V} = \frac{12g}{1cm^3} = 12\ g/cm^3$$

If volume increases → $V = 3cm^3 \quad D = \frac{M}{V} = \frac{12g}{3cm^3} = 4\ g/cm^3$

$V = 4cm^3 \quad D = \frac{M}{V} = \frac{12g}{4cm^3} = 3\ g/cm^3$

$V = 6cm^3 \quad D = \frac{M}{V} = \frac{12g}{6cm^3} = 2\ g/cm^3$

Types of Graphs

The relationship between two measurable factors can be represented by a graph. If both values increase, they are considered to have a direct relationship. For example, the more you work, the more money you make. To use a sports analogy, the more you practice, the more points you will score. If one value increases while the other decreases, they have an inverse relationship. An example of this is the more you work on your golf swing, the lower your score will be. Cyclic graphs show one variable alternating its value in an increasing and decreasing pattern while the other variable remains constant. Sunspots can be charted to show this pattern. The rising and falling level of the ocean tide is graphed by a cyclic graph. If one variable stays constant while the other increases, the graph shows a horizontal or vertical line. A graph plotting density and size would show this type of relationship. If there is no relationship that can be found between two variables, no line can be drawn on the graph. These graphs are in Figure 1-2.

Figure 1-2 Graphs showing relationships.

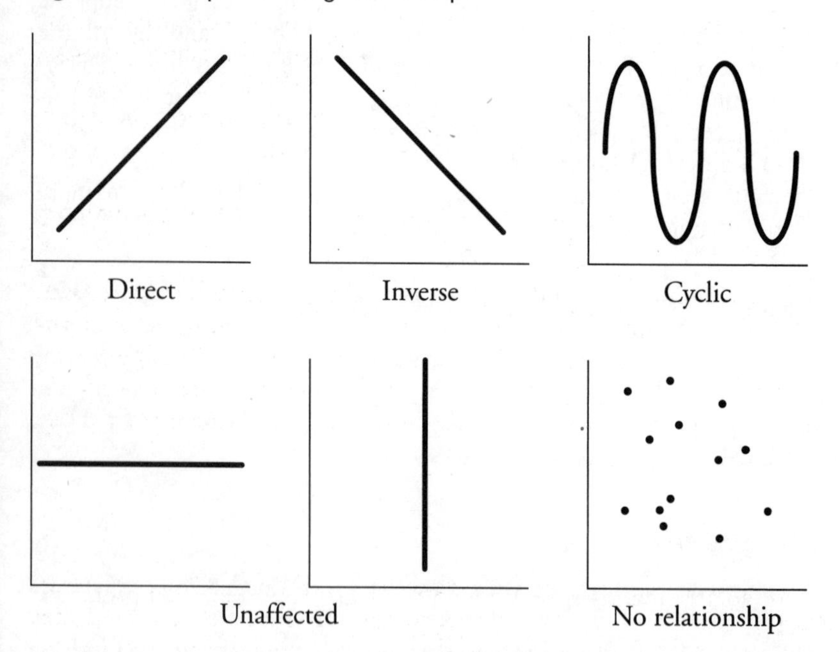

Several types of graphs can be used to show relationships. **Line graphs,** which were discussed in the preceding paragraph, are used extensively in Earth Science. Line graphs use a **coordinate system** to plot the points. The independent variable is plotted along the horizontal axis (*x*-axis) and the dependent variable is plotted along the vertical axis (*y*-axis). The line drawn can connect the points from dot to dot. A line of best fit can also be drawn. This is a straight line running through most of the points.

A pie chart can be used to organize data in a different way. This chart can show how something is divided into its component pieces. The percentages of the elements that make up the Earth's crust can be depicted clearly in a pie chart.

If one variable is being observed over time, it is sometimes useful to show this with a bar graph. The rainfall for an area can be represented with a bar graph, with a different bar representing each year's rainfall.

Natural Resources and Conservation

The amount of natural resources that we have on the Earth is limited. These are substances that come from the Earth. **Renewable** resources are replenished shortly after they are used. Examples of these are wind, solar, geothermal, and trees. If the resource cannot be replaced for millions of years or at all, it is **nonrenewable.** Fossil fuels (coal, oil, and natural gas), radioactive materials (for nuclear energy), groundwater, minerals and metals from the ground are all nonrenewable resources. Unless we change our ways, the finite amounts of these materials will be used long before they are replenished.

There are several methods that we can use to conserve these resources. Reducing the amount used will help stretch the remaining amount further. By reusing some materials, we can help ensure that the supply won't run out soon. Many areas are recycling their materials, which can help ease the need for new materials. This also decreases the amount of garbage that we make.

Pollution

The Earth is in a balanced state. In a very short time, geologically speaking, humans have disrupted this **equilibrium.** There are many areas that we have polluted. **Pollution** is a substance that harms living organisms or the environment. In some instances, the Earth can return to its original state of **dynamic equilibrium,** but in others it can take a long time, if at all, to repair the damage done. With the increase of industry and technology, we have sped up the pollution process in some instances.

Types of Pollution

Air pollution can be caused by many factors. Carbon dioxide, a **greenhouse gas,** can be made through the process of respiration and as a by-product of burning fossil fuels. Carbon dioxide, as well as other greenhouse gases such as methane and water vapor that are found in the atmosphere, can absorb longwave radiation coming from Earth and can hold this heat, which raises the air temperature.

Water pollution can be caused by many sources. These range from mercury levels that are harmful to humans, to fertilizers that are washed into the water supply.

The ground can also be polluted by human activities. The disposal of garbage and wastes can affect the ground. Chemicals that leach into the soil can be harmful to plants and animals.

Chapter Checkout

Q&A

1. Using a ruler to measure the length of a stick is an example of

a. extending the sense of sight by using an instrument.
b. calculating the percent of error by using a proportion.
c. measuring the rate of change of the stick by making inferences.
d. predicting the length of the stick by guessing.

2. The diameter of Jupiter through its equator is about 143,000 km. What is this distance written in scientific notation (powers of 10)?

a. 143×10^2 km
b. 1.43×10^3 km
c. 1.43×10^5 km
d. 143×10^5 km

3. A pebble has a mass of 35 g and a volume of 14 cm^3. What is its density?

a. 0.4 g/cm^3
b. 2.5 g/cm^3
c. 490 g/cm^3
d. 4.0 g/cm^3

4. A student measures his latitude to be 50.0° N when it is actually 40.0° N. What is the percent deviation (percentage of error) in his measurement?

a. –10%
b. –15%
c. –20%
d. –25%

Answers: 1. a **2.** c **3.** b **4.** d

Chapter 2

MEASUREMENTS AND MODELS OF THE EARTH

Chapter Checkin

- ❑ Knowing the Earth is extremely round and smooth
- ❑ Understanding how the Earth's circumference was calculated
- ❑ Determining the variety of the Earth's layers

A model of the Earth can show its shape and surface features with some accuracy. The Earth is actually a small to medium-sized planet when compared to the others in our solar system. The shape is nearly spherical, and it has a very smooth surface.

Size, Shape, and Roundness of the Earth

The measurement of the size and shape of the Earth has had a long history. Although the circumference was calculated long before sailors circumnavigated the Earth, newer technology has just precisely quantified what was already known.

The Size of the Earth

More than 2,000 years ago, a man named Eratosthenes calculated the circumference of the Earth with about 99 percent accuracy. He lived in Egypt. He used information he obtained in his dealings with travelers, a simple mathematical formula, and used his own genius to put this all together to figure out the circumference. He assumed that the Earth was perfectly round and that the rays from the sun that reach the Earth are parallel. He also knew from geometry that alternate interior angles are equal. He measured the shadow angle of a building in his city. On June 21, in a city due south of Eratosthenes, the sun was directly overhead at noon. There was a well in the center in the city, and people could see to the bottom of it on this day. By knowing the distances between the cities and the angle of the shadow, he was able to calculate the circumference of the Earth. Figure 2-1 is a diagram of this measurement and the formula that he used.

Figure 2-1 Eratosthenes equation and diagram.

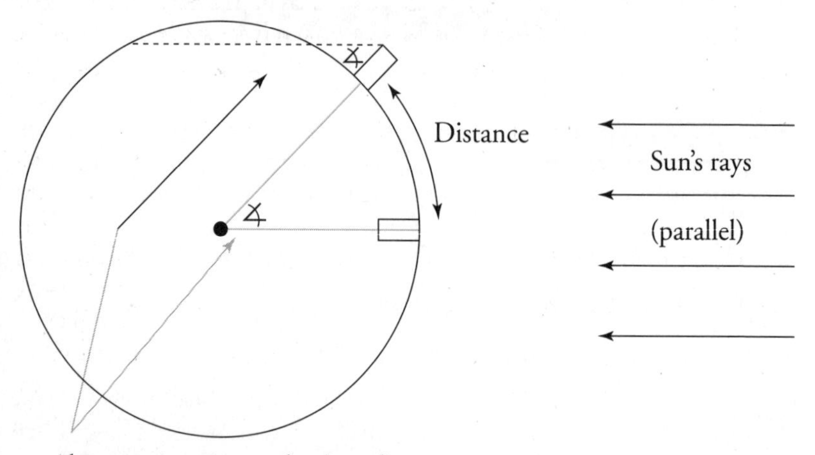

Alternate interior angles (equal)

$$\frac{\measuredangle^\circ}{360^\circ} = \frac{S}{C}$$

$\measuredangle^\circ$ = shadow angle
S = distance between building, well
C = circumference of the Earth

We know now that the equatorial circumference of the Earth is slightly larger than the polar circumference, which is the one that Eratosthenes calculated. This is still an incredible and incredibly accurate achievement without the use of modern technology.

The Shape of the Earth

When looked at from the moon, the Earth looks perfectly round. Upon further measurement, this is not exactly true. The polar diameter is smaller than the equatorial diameter by 42 km (about 25.6 mi). This might seem a long way to run; however, when you compare it to the actual diameters of the Earth's polar circumference (12,714 km) and equatorial circumference (12,756 km), this is a very small distance. The term **oblate spheroid** is used to describe this shape. The Earth is flatter at the poles and slightly bulging at the equator. The reason for the slight bulge at the equator is because of the centrifugal force created by the Earth's rotation. This is the same force that you feel pushing outward, away from the center of a turn, when you ride on a roller coaster. Because you are farther away from the center of the Earth at the equator (at sea level), you weigh less there than you would at the poles. It is a barely measurable amount, but it exists.

The idea that the Earth is round is not new. It has been around since at least the fifteenth century. We have had pictures taken from space and from the moon to help support the idea. There are many historical examples that show the idea that the Earth is round and not flat. When ships approached a city on the coast, the top of the mast was seen first. As ships got closer, the people in the city would gradually see more and more of the ship. This idea was also used to prepare for war. The inhabitants of the city knew how tall ships were and how fast they traveled. With some geometry and speed formulas, they could predict when the ships would arrive. Sailors also noticed that as they sailed north and south on the Earth, the constellations that they saw varied. Therefore, the view seen when you look "straight up" differs depending on your latitude. This added to the idea that the Earth was round. Although you cannot directly see the Earth curving, you can see it indirectly. As the full moon moves into the Earth's shadow during a lunar eclipse, the shadow of the Earth is observed on the moon.

The Smoothness of the Earth

If you were to stand in the middle of the Rocky Mountains or the Himalayan Mountains, the idea that the Earth is smooth wouldn't even appear in your mind. When the size of the whole Earth is compared to the heights of Mt. Everest and these mountain ranges, the smoothness ratio is very close to 1.0, or perfectly smooth. The roundness ratio of the Earth has very similar results. If the polar diameter is divided by the equatorial diameter, the result is very close to 1.0. This means that the Earth is an extremely round and smooth planet. It has been said that Kansas is flatter than a pancake. A study was done on this, and the results were published at the following Web site, confirming this idea: http://www.improbable.com/airchives/paperair/volume9/v9i3/kansas.html.

The Structure of the Earth's Layers

The Earth can be divided into several spheres. The outermost sphere is the gas layer, which is the least dense. This is the **atmosphere.** The atmosphere can be subdivided into layers that have specific properties. On the surface of the Earth is the **hydrosphere.** This is the liquid layer of water on the Earth consisting of the oceans, lakes, and rivers. The layer of rock found on the outer edge of the Earth is the **lithosphere.** This layer is divided into plates. The interior of the Earth, which contains the densest material, is divided into the mantle, inner core, and outer core.

Types of Models

Models are representations of other objects. Usually these objects are too large or too small to be seen with your eyes. The **scale** of a model represents the amount that the original proportions were changed. In other words, the scale is how much the object was shrunk or how much it was magnified.

There are several different types of models. Physical and mechanical models provide information through your sense of sight. Globes, model cars, and doll houses are good examples of **physical models.** If you add moving parts, they become **mechanical models.** Electric trains and radio-controlled cars fall into this category. Most objects start out as an idea. These are **mental models,** which exist in your mind. **Mathematical models** are also called equations. **Graphical models** use a picture to show a relationship. The type of data present and the reason for making the graph determine which type of graph will be used to best represent the information.

Chapter Checkout

Q&A

1. The true shape of the Earth is best described as a

a. very oblate spheroid.
b. perfect sphere.
c. perfect ellipse.
d. slightly oblate spheroid.

2. The polar circumference of the Earth is 40,008 km. What is the equatorial circumference?

a. 12,740 km
b. 25,000 km
c. 40,008 km
d. 40,076 km

3. In which group are the spheres of the Earth listed in order of increasing density?

a. atmosphere, hydrosphere, lithosphere
b. hydrosphere , lithosphere, atmosphere
c. lithosphere, hydrosphere, atmosphere
d. lithosphere, atmosphere, hydrosphere

4. The distance between two points on the same longitude is 2,000 km. The sun is overhead at one point, and the shadow of a building makes a 45-degree angle with the ground at the other point. What is the circumference of this planet?

a. 9,000 km
b. 12,000 km
c. 16,000 km
d. 24,000 km

Answers: 1. d **2.** d **3.** a **4.** c

Chapter 3
MAPPING THE EARTH

Chapter Checkin

- ❑ Understanding the latitude and longitude grid system on Earth
- ❑ Knowing the rules for making field and topographic maps
- ❑ Determining the gradient between two points on a topographic map

Making accurate maps of the Earth's surface has been a challenge since the earliest maps were made. Round objects such as the Earth are difficult to translate into flat maps. Distortion of the maps occurs, usually at the poles; therefore, different types of maps have been used throughout history to show the Earth's surface. Advances in technology have increased the accuracy of making maps while allowing more information to be presented on the maps.

Latitude and Longitude Measurements

An imaginary man-made grid, which helps us to locate a position on Earth, covers the Earth's surface. The coordinates of **latitude** and **longitude** have been developed to take into account the spherical shape of the Earth. The lines were developed to combine two different methods of measurement. One set of lines runs parallel to each other (these are latitude). The other set of lines are circles meeting at the poles, and they are farthest apart at the equator (these are longitudes). The two systems that are combined to make this grid are rectangular coordinates and polar coordinates.

Latitude

Latitude lines run east to west, and are parallel to each other. If you see them on a globe, they look like the rungs on a ladder. These **parallels** are measured in degrees north and south of the equator. The equator is labeled 0°

and this number increases in both directions towards the North and South poles, where the latitude for each is 90°. Because the lines are parallel, the distance between the lines is constant (1° = 70 mi or 112 km). The latitude lines are angles measured from the center of the Earth out to the surface, compared with a line from the center of the Earth out to the equator. This is illustrated in Figure 3-1.

Figure 3-1 Latitude lines and their measurement from the center of the Earth.

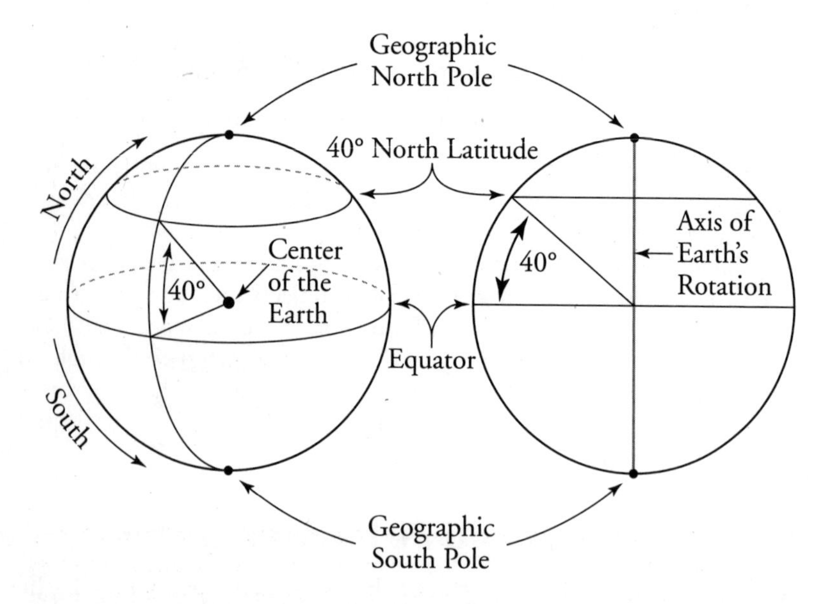

The latitude of an observer in the Northern Hemisphere can be determined by finding the altitude of **Polaris** (the North Star) above the horizon. If you are at the North Pole, Polaris, which is the tip of the handle in the Little Dipper constellation, is directly overhead. In other latitudes in the Northern Hemisphere, Polaris can be found by finding the Big Dipper and then following the "pointer stars" (the two at end of the ladle part) to Polaris.

Longitude

The lines of longitude run in a north-south direction, through the poles. These are the long lines on a globe and are the farthest apart at the equator. They are labeled in degrees east and west from the **Prime Meridian.** The

Prime Meridian is the reference point for longitude and runs through Greenwich, England. Each of these **meridians** is equal in length. Figure 3-2 shows where these angles are measured internally in the Earth.

Figure 3-2 Longitude lines and their measurement from the center of the Earth.

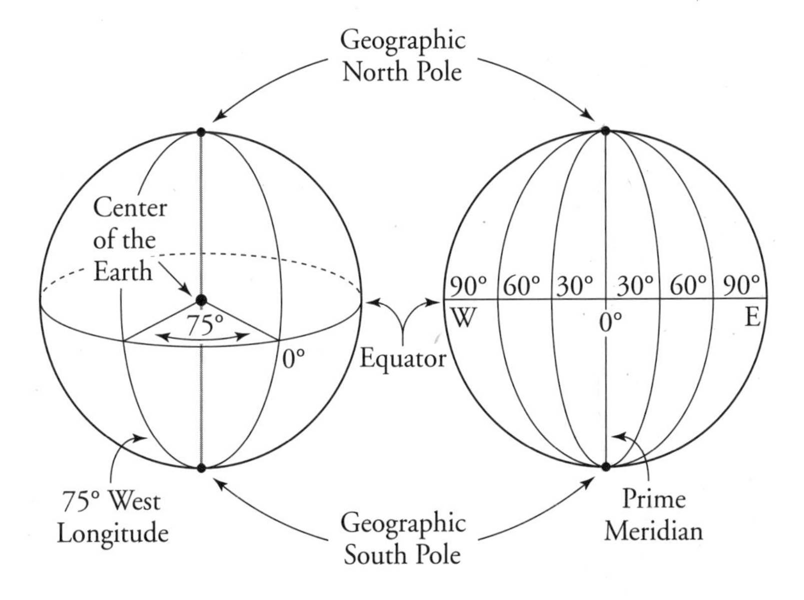

Because The Earth takes about 24 hours to rotate and since there are 360° of longitude, the Earth takes 1 hour to rotate 15°. This is how time zones are set up across the globe. On the opposite side of the Earth from the Prime Meridian is the **International Date Line,** which has a longitude of 180°. This is where the date switches (or "where the day begins"). **Local noon** or high noon occurs when the sun is directly on your meridian. The sun might not be directly overhead, but it is at its highest point for the day. If the sun hasn't reached its peak yet, this is morning, or A.M., which stands for ante meridian. The afternoon time, when the sun has passed its apex, is called P.M., or post meridian.

Great Circles

When planes fly around the globe, the shortest distance between the two points isn't always along what would be a straight line on a flat map. The best route for planes to follow is along great circles. The shortest distance between two points along the surface of a round object is usually along the curve of a great circle.

Field Maps

When measurements, such as elevation, temperature or pressure to name a few, are made across a given area, a map can be made of these readings. This is helpful because these values can then be analyzed and interpreted. A **field** is an area in which there is a measurable amount of a specific value at every point. Some of these fields can be **scalar** quantities. These are units measured in magnitude, such as temperature, pressure, and relative humidity. Other fields are expressed as **vector** quantities. These measurements are given in terms of both magnitude and direction. Wind velocity, gravity, and magnetism are measured as vector quantities. On field maps, lines connect equal values and these are called **isolines.** If you were to walk along these lines, the readings wouldn't change. These lines are commonly seen on weather maps that show temperature or pressure. On occasion, **isosurfaces** are drawn. These are 3-D diagrams showing surfaces that connect points with the same field value. A gradient can be calculated to figure out the rate at which the field quantity changes. This can also be thought of as the slope of the field. These generally change over time. The formula for finding the gradient is as follows:

$$\text{Gradient} = \frac{\text{amount of change in the field quantity}}{\text{distance over which change occurs}}$$

Topographic Maps

Field maps that show elevations in relation to sea level are called **topographic maps.** These maps show land forms and features. The U.S. Geological Survey maps are divided into quadrangles. They are maps that are usually 7.5' of latitude and 7.5' of longitude on each side. The scale of the map, which shows how much the map "shrinks" down the land, is 1:24,000. This means that 1" on the map is equal to 24,000" (2,000') on Earth.

Contour Lines

The lines connecting points of equal elevation are called **contour lines.** The **contour interval** is the difference between contour lines and generally is 10', 20', or 50', depending on the terrain. Every fifth line on a topographic map is thicker and labeled with the elevation, which is called the index contour.

This is done to help in counting the elevation on a map. Some maps have the contour interval written below the map, as part of the key. Contour lines are continuous on a map. Contour lines can also show the slope of the land. The lines can get close to each other to show an area that has a steep slope. And if the area has a gentle gradient, the contour lines are more spread out. If the area is very steep, such as a cliff, the lines may touch, but will never cross over each other. Contour lines that are closed onto themselves show mountains. As contour lines come across a river or stream, they bend upstream. This is because the land is not flat, and water moves downhill. As the contour lines reach a lake, they go around it because the surface of the lake is flat. Depression contours show where the land goes into a depression or an indentation on the Earth. These are found in swampy areas or in craters of a mountain. They are also shown with concentric circles, but are represented differently than regular contour lines by hachure marks, which can be seen at the bottom right in Figure 3-3.

Figure 3-3 Sample topographic map.

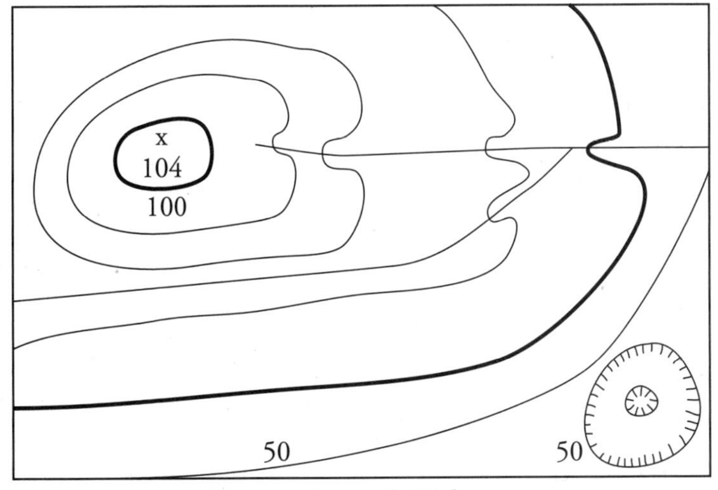

If the top of a hill or mountain, or some other location is accurately measured at a specific point in time, it will be shown on a topographic map with a benchmark. The symbol for this is an *X, BM,* or a $\triangle$, with the exact elevation written on the map. Otherwise, the highest possible elevation for a hill or a mountain is 1' below the next contour line.

Reading Topographic Maps

Topographic maps are drawn to **True North,** but have an angle at the bottom of the map or a compass rose that shows **magnetic declination.** This **Magnetic North** direction is different for each location, and for when the map was made. This varies slightly each year as the magnetic poles migrate. The slope (gradient) of the land can be found by changing the gradient equation:

$$\text{Slope} = \frac{\text{change in elevation}}{\text{distance over between the points}}$$

Profiles can be made from topographic maps. These are side views of the land's surface. Topographic maps show the land from a top view (bird's-eye). A profile is drawn from one given point to another. The path you would take connecting these two points is shown on a graph. You would make a dot on a piece of paper to represent every time you cross a contour line. You would then make dots at the proper elevation on a different piece of paper, and connect those dots to show what the profile of the cross section would look like. How far you space those elevations on the *y*-axis will determine the scale of the profile (which may not be the same as the horizontal scale). Figure 3-4 shows an example of a profile.

Figure 3-4 Profile of a topographic map.

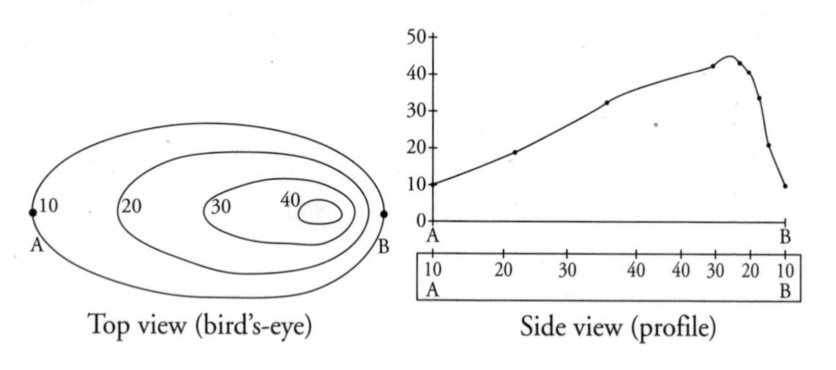

Updating Maps

The original topographic maps were made by people actually going into the field and surveying the land. These maps were revised over time using photographs that were taken from planes flying overhead. Roads and buildings were added during these updates. Light and sound waves emitted from planes were used to help determine land elevations. More recently, satellites have been used to take images of the land to make and update topographic maps.

Chapter Checkout

Q&A

1. A vector quantity must include both magnitude and direction. Which measurement is a vector quantity?
   a. The motion of water in an ocean current
   b. The air temperature in a room
   c. The rain accumulation at a weather station
   d. The highest elevation of a hill
2. Which statement is true about an isoline on an air-temperature field map?
   a. It indicates the direction of maximum insolation.
   b. It represents an interface between high and low barometric pressures.
   c. It connects points of equal air temperature.
   d. It increases in magnitude as it bends southward.
3. A contour map shows two locations, *X* and *Y*, 5 km apart. The elevation at location *X* is 800 m, and the elevation at location *Y* is 600 m. What is the gradient between the two points?
   a. 12 m/km
   b. 40 m/km
   c. 120 m/km
   d. 160 m/km

4. Isolines on the map show elevations above sea level, measured in meters. What is the highest possible elevation represented on this map?

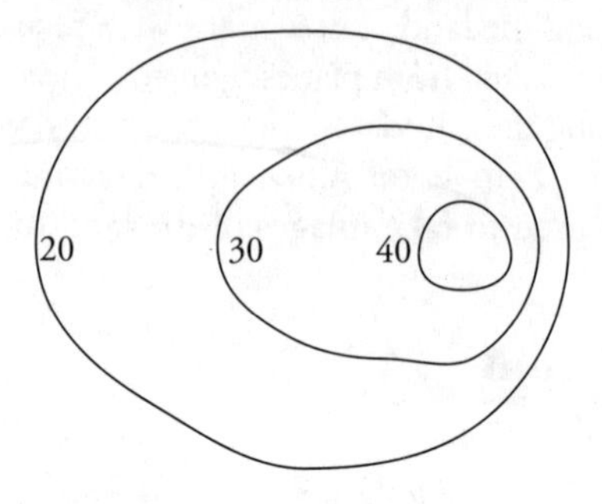

a. 39 m
b. 41 m
c. 49 m
d. 51 m

Answers: 1. a **2.** c **3.** b **4.** c

Chapter 4
MINERALS

Chapter Checkin

- ❑ Understanding what minerals are and how they are made
- ❑ Identifying a mineral by its properties

Minerals are all around us and are made up of matter. **Matter** is anything that has mass and takes up space. Minerals are made up of elements and molecules. Elements are substances that can't be broken down into simpler substances by chemical means. In the Earth, there are 92 naturally occurring elements. These combine together to make up more than 3,000 different kinds of minerals. However, only eight elements (oxygen, silicon, aluminum, iron, calcium, sodium, potassium, and magnesium) make up 98.5 percent of the total mass in the Earth's crust. Most of these are found in **compounds,** which are made up of two or more elements. Examples of compounds are quartz (SiO_4), halite ($NaCl$), and galena (PbS). Some minerals occur as single elements; examples include gold, silver, copper, sulfur, and diamond. Minerals are formed from magma cooling or as evaporites in water. The organization of the atoms in a mineral give it physical and chemical properties.

Crystal Structure

Minerals are naturally occurring, inorganic materials that are solid. They have definite compositions and have definite properties that will be discussed later in this chapter. Minerals are crystalline, which means the atoms are arranged in an orderly pattern that repeats throughout the mineral. Several crystalline structures appear more often than others; these structures are shown in Table 4-1.

Table 4-1 Common Crystalline Structures

Type of Structure	*Diagram of Crystal*	*Examples*
Tetrahedron	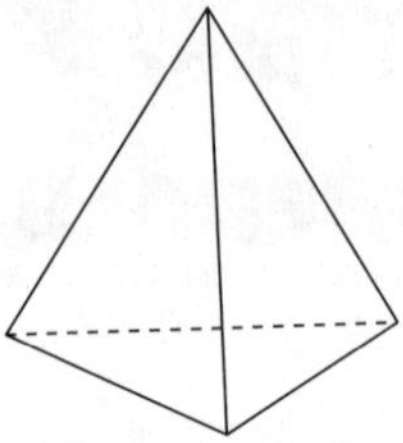 Silicon tetrahedron	Quartz
Cubic	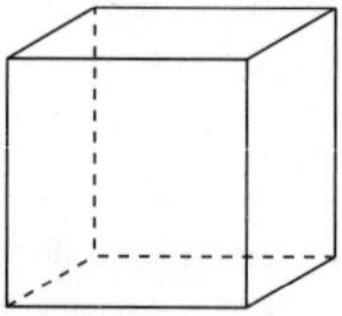 Cubic	Halite, galena
Hexagonal	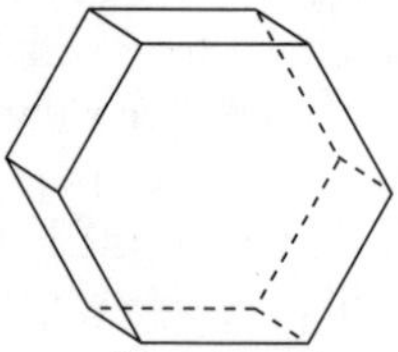 Hexagonal	Graphite

The family of tetrahedrons can be further broken down into several groups. If the tetrahedron is a single molecule and attached to other tetrahedrons by other elements, it is called ionic. Olivine is an example of this type of mineral. The molecules could be linked together in a single chain, which would be represented by pyroxene. Amphibole is an example of a mineral that has a double chain of molecules. The molecules could also be arranged in a sheet. Mica is a great representation of this. Quartz is a mineral that has all of the tetrahedra connected in a network, which is part of why it is so strong.

Mineral Identification

Minerals can be identified through a series of criteria. Some of these are done visually and others are done through simple tests.

The visual characteristics of minerals are as follows:

- **Color:** Although this is the first characteristic that we can observe, it isn't always the most reliable. Impurities can change a mineral's color dramatically. Quartz has 33 different varieties of color, but sulfur is always a bright yellow.
- **Luster:** This is the way a mineral shines in reflected light. Minerals are metallic or nonmetallic. Metallic minerals include galena, graphite, and pyrite. Graphite is the "lead" in pencils. Pyrite is also known as "fool's gold." Nonmetallic minerals don't have a shiny metal-like sheen. If a mineral looks glassy (vitreous), as does quartz, it is considered to have a nonmetallic luster. Other types of nonmetallic luster include pearly, silky/fibrous, greasy/waxy/resinous, and earthy/dull.
- **Crystal shape:** The arrangement of the molecules in the mineral leads to the shape of the crystals that we see. Halite (table salt) and galena are seen as cubes. The crystal structure of fluorite is two pyramids stuck together at the bases, making an octahedron.

The simple tests for identifying minerals are as follows:

- **Streak:** This is the color of the powder left behind when a mineral is rubbed on an unglazed porcelain tile. The streak might not be the same color as the mineral.
- **Cleavage:** The way that a mineral splits or breaks along weak bonds between the molecules is its **cleavage.** Mica splits easily in one direction, as is seen in its sheeting action. Feldspar can readily split in two directions. The cubic nature of galena and calcite show that these minerals split in three different directions.
- **Fracture:** Minerals that break unevenly are considered to fracture. One specific example of fracture is obsidian. The curved circles along the break are called conchoidal fracture.
- **Hardness:** Minerals are classified by whether they can scratch each other. Mohs hardness scale is a relative scale measuring this and has a range of 1 to 10. This scale is shown in Table 4-2.

Table 4-2 Mohs Hardness Scale

Mineral	*Hardness*	*Simple Test*
Talc	1	Fingernail can scratch it.
Gypsum	2	Fingernail can scratch it.
Calcite	3	Fingernail can scratch it.
Fluorite	4	Can't scratch glass.
Apatite	5	Can't scratch glass.
Feldspar	6	Scratches glass.
Quartz	7	Scratches glass.
Topaz	8	Scratches glass.
Corundum	9	Scratches glass.
Diamond	10	Scratches glass.

- **Specific gravity:** This is the relative density of a mineral compared to the density of water. In other terms, it is how heavy a mineral feels when you pick it up.
- **Acid test:** Some minerals that contain calcium carbonate will fizz when acid is put on them. Calcite, which is the main component of limestone and marble, will react this way to acid rain.

Some minerals have special properties that are specific to them and help us to clearly identify them. These include: magnetism (magnetite), salty taste (halite), fluorescent glow under UV light (fluorite, calcite), rotten-egg smell (sulfur), double image (Icelandic Spar—a form of calcite), phosphorescent glow after light is turned off (willemite, sphalerite), and radioactivity (carnotite, uraninite).

Mineral Families

Silicates are minerals that are made from silicon and oxygen and comprise most of the Earth's crust. These are composed of silica tetrahedrons and include quartz, feldspar, mica, talc, amphibole (hornblende), pyroxene, olivine, garnet, and kaolin.

Minerals made from carbon and oxygen are called **carbonates.** These minerals react to acid and include calcite, dolomite, malachite, and azurite.

The family of iron oxides is comprised of minerals that are made from iron, sulfur, and oxygen. These minerals can appear to be rusty. Examples of minerals in the iron oxide family are hematite, magnetite, and pyrite.

Chapter Checkout

Q&A

1. The mineral mica breaks evenly along flat sheets mainly because of its

a. atomic arrangement.
b. chemical composition.
c. hardness.
d. density.

2. The relative hardness of a mineral can best be tested by

a. breaking the mineral with a hammer.
b. determining the density of the mineral.
c. squeezing the mineral with calibrated pliers.
d. scratching the mineral across a glass plate.

3. Which mineral fizzes when dilute hydrochloric acid (HCl) is placed on it?

a. Calcite
b. Feldspar
c. Quartz
d. Talc

4. Which property is illustrated by the peeling of muscovite mica into flat, thin sheets?

a. Luster
b. Streak
c. Hardness
d. Cleavage

Answers: 1. a **2.** d **3.** a **4.** d

Chapter 5
ROCKS

Chapter Checkin

- ❑ Understanding the different ways that rocks are formed
- ❑ Determining how a rock should be classified
- ❑ Knowing how a rock can be transformed into another rock

Rocks make up the solid part, or lithosphere, of the Earth. There are three different types of rock: igneous, sedimentary, and metamorphic. The basic difference between the three different types of rocks is how they were formed. Rocks are generally made from minerals. The same materials have been used over and over again since the Earth was formed. The theory of Uniformitarianism was proposed in 1795 by James Hutton. His theory states that processes that occurred in the past are happening today, and that the physical features of today took a long time to occur. Prior to that, the general assumption by scientists was that these things occurred quickly as catastrophic events.

Igneous Rocks

Igneous rocks were the original rocks formed while the Earth cooled. The word *igneous* means "fire-formed." When a rock is melted and allowed to cool, it becomes an igneous rock.

You can identify an igneous rock by two main characteristics: texture and composition. Texture can also be thought of as the size of its grains or crystals.

Texture

If **magma** reaches the surface of the Earth, its chemistry changes as it interacts with the atmosphere. The liquid rock is called **lava.** Lava cools quickly and is called an **extrusive** rock. The crystals that form are small. Sometimes, the rock cools so quickly that no crystals form, and this is called volcanic glass. Lava flows generally come from volcanoes. A way to remember this is that Vulcan is the Roman god of fire.

If magma stays below the surface of the Earth, it will cool more slowly, which allows for larger crystal growth. Therefore, the longer the cooling period is, the larger the crystals are. These rocks are called **intrusive** or **plutonic** rocks. They are named for Pluto, the Roman god of the underworld.

Composition

The composition of magma can vary from being **felsic** to being **mafic.** Felsic magmas are aluminum-based and are lighter in color. They also have a lower density. Mafic magmas are iron and/or magnesium based. These are darker in color and have higher densities.

Other materials present also contribute to the color and density of the rock. Felsic rocks are generally made from potassium feldspar, quartz, plagioclase feldspar, biotite mica, and amphibole (hornblende). Mafic rocks are usually made from olivine, pyroxene, plagioclase feldspar, amphibole, and biotite mica. The amounts of each mineral can vary from rock to rock.

Occasionally, you might find a rock with large crystals that are surrounded by smaller crystals. The crystal formations happened at different times. This type of igneous rock is considered to have a **porphyritic** texture.

In Table 5-1, you can see the how igneous rocks are organized. They are placed in the table according to the size of their grains and their composition. You can divide the rocks in the table into three types: mafic, felsic, and those that fall in between.

Table 5-1 Scheme for Igneous Rock Identification

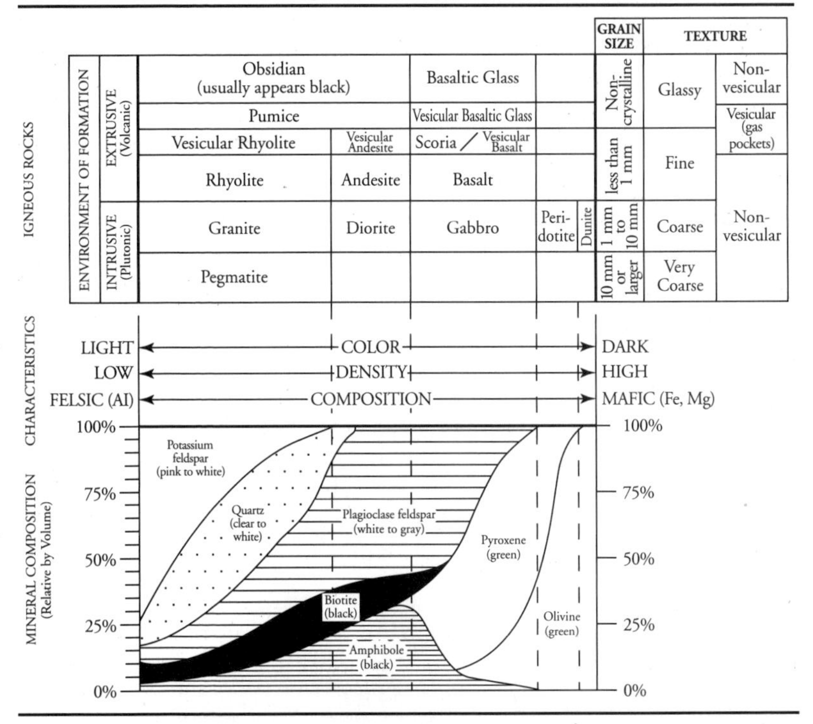

The most common igneous rock found in the continents is granite. Under the oceans, you will generally find basalt. One way to remember that basalt is found under the oceans is to think that the oceans are "salty." Some igneous rocks that are formed from volcanoes cool so quickly that they trap air in the rocks. Pumice is an example of this. A rock is considered to have a **vesicular** texture if it has gas pockets. Obsidian is an igneous rock that cools so quickly that it looks like glass and appears black. It is actually felsic in composition, even though it is dark in color.

Sedimentary Rocks

When rocks on the surface of the Earth are weathered, eroded, and deposited, they can form **sedimentary** rocks. Some of these rocks are formed from inorganic fragments of other rocks. Other sedimentary rocks can be chemically or organically formed. The particles are weathered and transported by running water, wind, waves, and glaciers.

Clastic rocks (*clastic* is Greek for *broken*) are pieces of other rocks that are cemented together by silica, lime, and iron oxide. This usually occurs under water. The pressure from water and layers of sediment above cements together these transported fragments. The name of the rock comes from the size, or the range in sizes, of the grains that have been cemented together.

Sedimentary rocks generally are made of the same size grain due to the deposition of eroded sediments. As a river enters an ocean or lake, it slows down. The particles that were transported by the river are then deposited. They are separated by size and eventually will make a new sedimentary rock. Chapter 7, "Deposition," goes into more detail about this.

The chemically derived rocks come from crystals that form when water rich in minerals is allowed to evaporate. The crystals form in two ways. One way is that they can precipitate out when the concentration reaches its saturation point. The other method is if the water evaporates fully and leaves behind the crystals. These are called evaporites. Organic particles can also be formed into sedimentary rocks. Remains of animals, especially shells, can form limestone.

In Table 5-2, you can see how sedimentary rocks are classified.

Table 5-2 Scheme for Sedimentary Rock Identification

INORGANIC LAND-DERIVED SEDIMENTARY ROCKS					
TEXTURE	GRAIN SIZE	COMPOSITION	COMMENTS	ROCK NAME	MAP SYMBOL
Clastic (fragmental)	Pebbles, cobbles, and/or boulders embedded in sand, silt, and/or clay	Mostly quartz, feldspar, and clay minerals; may contain fragments of other rocks and minerals	Rounded fragments	Conglomerate	
			Angular fragments	Breccia	
	Sand (0.2 to 0.006 cm)		Fine to coarse	Sandstone	
	Silt (0.006 to 0.0004 cm)		Very fine grain	Siltstone	
	Clay (less than 0.0004 cm)		Compact; may split easily	Shale	
CHEMICALLY AND/OR ORGANICALLY FORMED SEDIMENTARY ROCKS					
TEXTURE	GRAIN SIZE	COMPOSITION	COMMENTS	ROCK NAME	MAP SYMBOL
Crystalline	Varied	Halite	Crystals from chemical precipitates and evaporites	Rock Salt	
	Varied	Gypsum		Rock Gypsum	
	Varied	Dolomite		Dolostone	
Bioclastic	Microscopic to coarse	Calcite	Cemented shell fragments or precipitates of biologic origin	Limestone	
	Varied	Carbon	From plant remains	Coal	

Sedimentary rocks can be identified by several characteristics:

- They are generally composed of particles with a similar grain size.
- They can show some layering. This can tell you about the history of the area, specifically how much water was there, how fast it was moving, and sometimes which direction the wind or water was going.
- The appearance of fossils occurs mostly in sedimentary rocks. These can include bones, impressions of shells or leaves or other such delicate details, which are usually destroyed during the formation of igneous and metamorphic rocks.
- Shallow-water ripple marks and mud cracks can be captured in the rock bed as well.
- Geodes can form in some instances. If water filled with minerals fills up a hole and then evaporates, a geode can form. Geodes are rocks that have cavities filled with quartz or calcite crystals.

Metamorphic Rocks

The third class of rocks is the **metamorphic** rocks, which are existing rocks that have undergone a great amount of heat and/or pressure. Sedimentary, igneous, or even other metamorphic rocks can be changed into a new metamorphic rock. The rocks never reach the melting point, but the crystals and fragments are rearranged and reformed. Intense heat is created from the friction of rocks moving and rubbing. The increased pressure is created from the weight of the rocks above or from the expansion or heat of nearby rocks that have become molten.

If a metamorphic rock's mineral crystals are layered, the rock is considered to have a **foliated** texture. Otherwise, metamorphic rocks are considered to be **nonfoliated.** Within each of these subdivisions, metamorphic rocks are separated by grain size and composition.

Regionally metamorphic rocks undergo intense heat and pressure over large areas. If hot magma forces its way into the rock above it, as it expands and gets less dense, it can "bake" the rocks with which it comes in contact. This is called **contact metamorphism.**

Table 5-3 shows how metamorphic rocks are identified.

Table 5-3 Scheme for Metamorphic Rock Identification

TEXTURE		GRAIN SIZE	COMPOSITION	TYPE OF METAMORPHISM	COMMENTS	ROCK NAME	MAP SYMBOL
FOLIATED	MINERAL ALIGNMENT	Fine	MICA	Regional	Low-grade metamorphism of shale	Slate	
FOLIATED	MINERAL ALIGNMENT	Fine to medium	MICA, QUARTZ, FELDSPAR, AMPHIBOLE, GARMET	(Heat and pressure increase with depth)	Foliation surfaces shiny from microscopic mica crystals	Phyllite	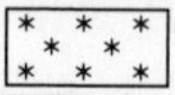
FOLIATED	MINERAL ALIGNMENT	Fine to medium	MICA, QUARTZ, FELDSPAR, AMPHIBOLE, GARMET	(Heat and pressure increase with depth)	Platy mica crystals visible from metamorphism of clay or feldspars	Schist	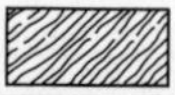
FOLIATED	BAND-ING	Medium to coarse	MICA, QUARTZ, FELDSPAR, AMPHIBOLE, GARMET, PYROXENE		High-grade metamorphism; some mica changed to feldspar; segregated by mineral type into bands	Gneiss	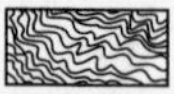
NONFOLIATED		Fine	Variable	Contact (Heat)	Various rocks changed by heat from nearby magma/lava	Hornfels	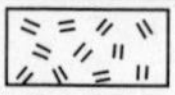
NONFOLIATED		Fine to coarse	Quartz	Regional or Contact	Metamorphism of quartz sandstone	Quartzite	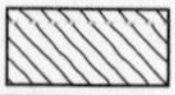
NONFOLIATED		Fine to coarse	Calcite and/or dolomite	Regional or Contact	Metamorphism of limestone or dolostone	Marble	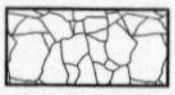
NONFOLIATED		Coarse	Various minerals in particles and matrix	Regional or Contact	Pebbles may be distorted or stretched	Metaconglomerate	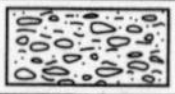

Some metamorphic rocks have great uses. The slate on chalkboards, pool tables, and roof tiles comes from shale, which has undergone metamorphic processes. Limestone that has been metamorphosed turns into marble, which is used for many purposes, such as headstones, countertops, statues, and flooring.

Metamorphic rocks can be identified by the distortion and/or twisting of the rock layers. Banding, or zoning of like minerals, can occur and results in the alternation of light and dark bands, which are parallel. These rocks are less porous and show an increased density. Chemical changes can also be noted.

The Rock Cycle

As noted earlier in this chapter, rocks and materials are constantly being recycled by the Earth. The original rock formed was igneous, but since the initial formation of the Earth, sedimentary and metamorphic rocks have been formed. There is no start or end to the **rock cycle**, but rather a continuous flow from one type of rock to another. There are many paths that link the different types of rocks. See Figure 5-1 for more details on these paths and the processes that occur.

Figure 5-1 The rock cycle in the Earth's crust.

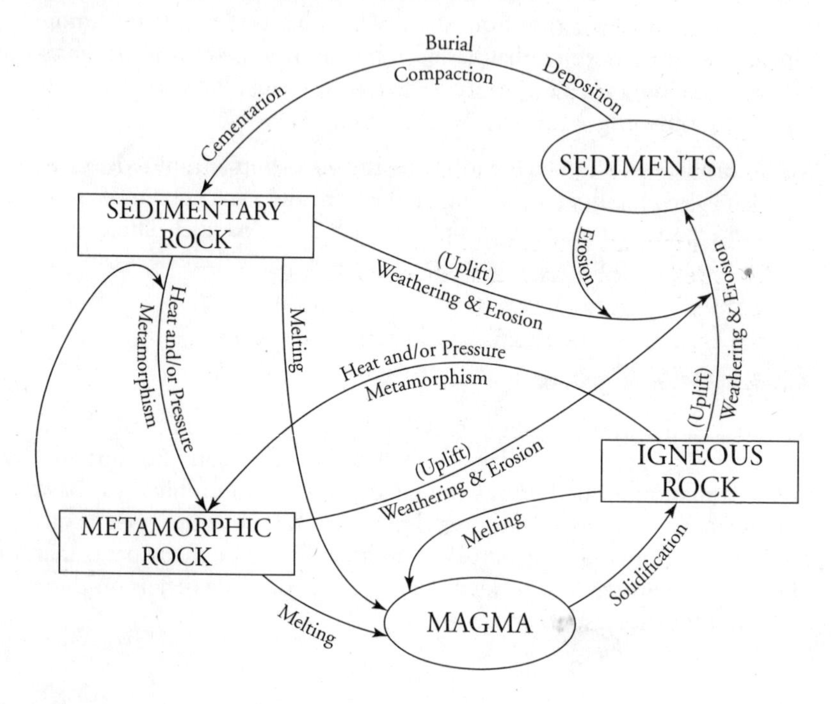

Chapter Checkout

Q&A

1. Which group lists rocks in order by grain size from smallest to largest?
   - a. Conglomerate, sandstone, shale
   - b. Sandstone, shale, conglomerate
   - c. Shale, sandstone, conglomerate
   - d. Shale, conglomerate, sandstone

2. Which minerals are present in granite but are never present in gabbro?
   - a. Quartz and plagioclase feldspar
   - b. Potassium feldspar (orthoclase) and quartz
   - c. Plagioclase feldspar and potassium feldspar (orthoclase)
   - d. Biotite mica and hornblende amphibole

3. Metamorphic rocks result from the

a. erosion of rocks.
b. recrystallization of rocks.
c. cooling and solidification of molten magma.
d. compression and cementation of soil particles.

4. According to Figure 5-1, which type(s) of rock can be the source of deposited sediments?

a. Igneous and metamorphic rocks only
b. Metamorphic and sedimentary rocks only
c. Sedimentary rocks only
d. Igneous, metamorphic, and sedimentary rocks

Answers: 1. c **2.** b **3.** b **4.** d

Chapter 6
WEATHERING AND EROSION

Chapter Checkin

- ❑ Understanding how rocks are weathered chemically and physically
- ❑ Determining how sediments are transported

The breaking down of rocks due to exposure to the atmosphere is known as **weathering.** There are two main categories of weathering—physical (also known as mechanical) and chemical. After rock is broken down, the particles and fragments are transported to another location by wind, water, or ice. This is known as **erosion.** These ideas are explained in more detail later in this chapter.

Types of Weathering

The physical breaking down of rock is called **mechanical weathering,** which can occur in several ways. These are dependent on the climate (temperature and precipitation) in an area. The main types of mechanical weathering are as follows:

- **Frost action:** This is also known as ice wedging and is caused by the freezing and thawing of water in a crack in the rock. As water freezes it expands, making the crack larger. After the ice melts, more water can fill up the crack. When this water freezes, the crack is again made larger. This keeps repeating until the rock breaks apart.
- **Plant action:** The growth of plant stems, trunks, and roots through cracks in rock can enlarge the cracks. This can often be seen as weeds grow up through the cracks of sidewalks and roads.
- **Exfoliation:** The top layer of rock is peeled off. Glacial advances and retreats can cause the top layer of rock to be removed.

- **Pressure unloading:** As the top layer of rock is removed, the bottom layers expand and crack. This is due to a decrease in pressure from the removal of the top layers. As the weight lessens, the rocks expand.

Rocks can also be weathered chemically. The rock material is broken down and changed into new substances. The same chemical changes that occur in chemistry class apply to **chemical weathering.**

- **Oxidation:** This occurs from the *rusting* of iron. You can see these reddish-brown streaks in magnetite, pyrite, and hematite.
- **CO_2:** As water dissolves calcite, carbonic acid is formed. This could eventually lead to limestone caves forming underground.
- **Acid rain:** Raindrops forming around molecules of SO_x and NO_x form acid rain. This dissolves carbonate-containing materials and lowers the pH of lakes.
- **Lichens:** Acids from the growth of lichens on rocks help to chemically break down rocks.

Rates of Weathering

The rate at which weathering occurs can be affected by several factors:

- **Climate:** The temperature and amount of moisture of a region can increase the rate at which weathering occurs. In areas prone to repeated freezing and thawing conditions, frost action happens. If the area is very warm and humid, it leads to more chemical weathering.
- **Particle size:** As the surface area of a particle increases, the rate of weathering increases.
- **Mineral composition:** The material that a rock is made from plays a large role in the rate at which the rock is weathered. Calcite-based rocks are more easily worn down than quartz-based rocks. This can create some magnificent geologic structures.

Soils

One of the products of weathering is the creation of soil. The combination of dead, decaying plants and animals with weathered sediments (organic and inorganic materials) makes up soil. Soil is created over time as the solid bedrock is exposed to weathering forces. As the bedrock is broken down and the area is inhabited by living things/different organisms, soils are created.

Soil Horizons

Soil can be broken down into different horizon layers, as shown in Figure 6-1.

- **Horizon A:** Topsoil (humus) is the rich, dark soil at the surface.
- **Horizon B:** Subsoil is more sandy than dark and has fewer nutrients.
- **Horizon C:** This is partially weathered bedrock.
- **Horizon D:** Bedrock is the solid layer of rock.

Figure 6-1 Four different soil horizon layers.

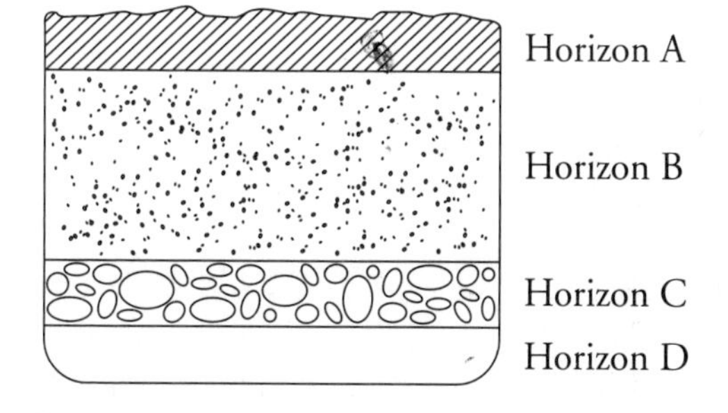

Soil Types/Climate

The type of soil that a region has depends on the climate and the type of plant growth. If the climate is warmer and wetter, decomposition occurs at a greater rate. You might think that tropical rain forests have deep, rich soil. The reality is that the soil is being used almost as quickly as it being made and the large amounts of rainfall leach out the rest. Here are some examples of climate conditions and typical soils:

- **Tropical soil:** Little minerals; thin layer; warm, wet conditions
- **Grassland soil:** Roots grow deep, thick topsoil layer, good for growing crops

- **Forest soil:** Acidic soil, rich in aluminum and iron
- **Desert soil:** Poor soil, high in calcium
- **Arctic soil**: Thin soil, poorly drained, boggy, permafrost present

Soil Conservation

With all of the forces trying to remove soil, several measures can be taken to prevent soil loss. Windbreaks can be made. Surrounding fields with trees creates a windbreak. Farmers can implement several techniques to help conserve their soil. The practice of contour farming can help. This method requires farmers to plow along the contour lines of the land. The alternation of the rows of crops, known as strip cropping, helps to prevent soil erosion. The process by which the field isn't disturbed between planting and the harvest is called no-till farming.

Agents of Erosion

After the bedrock material is weathered into smaller pieces, several agents of erosion transport the rock particles to another location. The primary methods for the transport to occur include running water, ice, wind, and waves. The overall driving force behind each of these agents is gravity.

Running Water

The greatest transporter of materials on the surface of the Earth is running water. A stream erodes the sides of its bed through the process of **abrasion.** As the water moves particles along, some particles are carried in the water column. These are known as suspension or colloid particles. Larger particles are bounced or rolled along the bottom, leading to rounding (or abrasion) of these particles. Chemical weathering can also occur. As the speed of the water increases, the size of the particle that can be carried by the running water increases. The amount of water passing a specific point in a given amount of time is known as **discharge.** This means that it is possible that wide, shallow rivers can carry the same amount of water as narrow, deep rivers. The speed of the river is also a factor in determining the discharge of a river. Table 6-1 shows the relationship between particle size transported by a stream and the speed of the stream.

Table 6-1 Relationship of Transported Particle Size to Water Velocity*

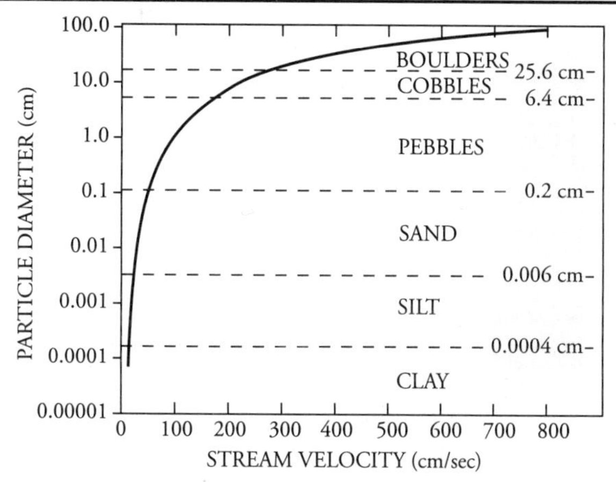

*This generalized graph shows the water velocity needed to maintain, but not start, movement. Variations occur due to differences in particle density and shape.

Velocity of a Stream

The speed of a stream at a given point can vary within the water column. Friction with the stream bed and air at the surface can slow down the velocity of the water. If the stream is moving straight, and you are looking down at it (top view), the fastest section is in the middle, away from the river banks. If you were to view the same stream from the side, the fastest point would be just below the surface (see Figure 6-2).

Figure 6-2 Views of a stream.

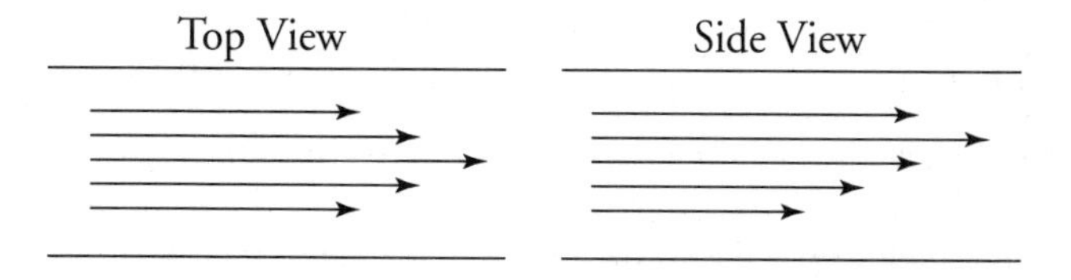

The fastest point of a stream differs for a meandering stream (see Figure 6-3). As a stream bends, the outside edge has to go faster to keep up with the water moving on the inside edge. The change in speed causes an area of erosion on the outer edge of the stream, while an area of deposition forms inside the curve. Over time, the differences in the speed of the stream and the erosional/depositional system created allow the stream to meander even more. This can bend the stream so much, it can sometimes lead to loops being cut off from the main flow, creating **oxbow lakes.**

Figure 6-3 A meandering stream's erosion and deposition.

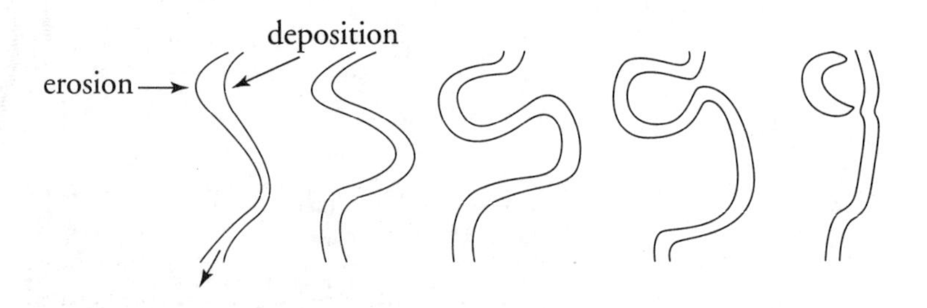

The Life of a Stream

In general, river valleys are characterized by a V-shaped valley. Glaciers, which are discussed in the next section, are identified by a U-shaped valley. The easy way to remember the difference is that rivers have the letter V in the word. The life cycle of a stream can be broken down into three stages:

- **Youth:** These rivers have steep sides and are fast moving. Lots of erosion occurs. The main direction of the erosional forces is in a downward direction (known as **downcutting**). Waterfalls and rapids are present during this stage.
- **Maturity:** As the stream gets older, the mountains that are being eroded get smaller. The gradient of the land gets lower, which means that the stream has less energy. This slower-moving stream starts to meander. The erosional forces start eroding laterally. The river valley widens.
- **Old Age:** The last stage in the life of a stream. This stage occurs when the gradient of the land is very low and flood plains are created. The stream bends back and forth to such a great degree that oxbow lakes are formed. These are lakes that were bends in a stream and were cut off from the main flow.

The stages of the life of a stream can be seen in Figure 6-4.

Figure 6-4 Life stages of a stream.

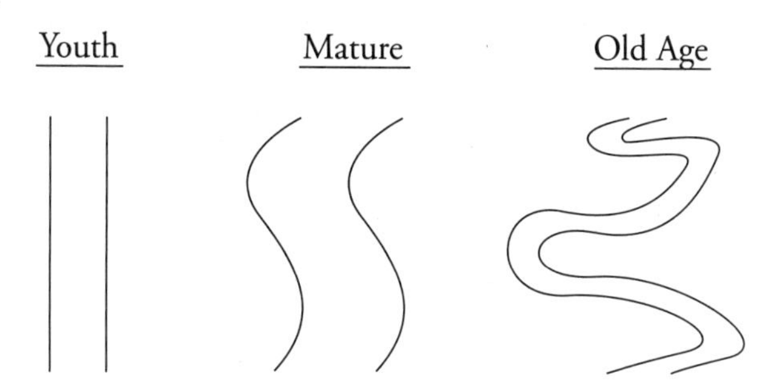

Glaciers

As ice and snow accumulate, compaction and recrystallization of the snow can occur. If the amount of snow and ice present grows faster than it melts and moves downward due to gravity, a glacier is formed. Glaciers are large areas of ice that covered about one-third of the Earth during the last glacial period. This period ended about 10,000 years ago. Currently, the Earth is in an interglacial period, or a natural period of warming. About 10 percent of the Earth's surface is now covered by glaciers. Many areas of North America were carved out by the glaciers that recently (geologically speaking) retreated. The Finger Lakes, the Hudson Valley and Long Island in New York, the Great Lakes, the soils of the Great Plains, and Yosemite Valley in California were all created by the Laurentide Ice Sheet that covered these areas.

Types of Glaciers

Glaciers are found in mountain areas and polar regions of the Earth. About 85 percent of the glaciers found today are located in Antarctica, and 10 percent are found in Greenland. If the glacial ice melted, the sea level would rise about 200 ft (60 m). Many cities along the coasts would be under water.

A **valley glacier,** also known as an alpine glacier, is located in a mountainous area and flows downhill through valleys. Most of the glaciers in North America are of this type.

A large area covered by a thick sheet of ice is considered to be a **continental glacier.** Currently these are found only in Greenland and Antarctica.

Glacier Formation

As snowflakes accumulate, the crystals on the bottom of the pile are compacted by the weight above. The points of the crystals are broken off, and the snowflakes are reformed. This is called firn or neve. This is also seen as "corn snow" by spring skiers. As the pile of snow gets larger, the pressure increases, which recrystallizes the flakes even further. The force of gravity, which had been pulling the ice and snow downward, now starts to pull the mass of snow and ice outward. This causes the glacier to move downhill. Icebergs are blocks of this ice that break off when the glacier reaches the ocean, which is called calving.

Glacial Movement

As glaciers move downhill, they are said to be advancing. If they melt faster than they are being created, they are in a period of retreating, or recession. The thickness of the ice and the gradient (steepness) of the slope affect the speed at which glaciers move. As with streams, the speed varies in different places of the glacier. Glaciers move more quickly in the center than at the sides and more quickly at the surface than at the bottom levels. The friction between the glacier and the ground is the cause for this discrepancy in velocity. The surface of the glacier is brittle. Fissures or cracks can form across the glacier as it moves. These are known as crevasses. The deeper area of the glacier is more plastic-like and flows in a smoother fashion. The upper part of the glacier, where more snow falls than melts away, is called the **Zone of Accumulation.** At the lower edge of the glacier, more snow melts than falls. The line between these areas, where snow is always present, is called the snow line. The water that melts off from the glacier is called meltwater.

Erosion by Glaciers

The movement of glaciers over the surface of the Earth causes characteristic features to be made. The surface of the Earth is scratched or abraded. Any meltwater that fills these cracks and scratches can freeze. This can loosen the bedrock through the process of frost action. If rocks are dragged along the bottom of the glacier, long scratches called **striations** are made. The orientation of these grooves in the rock can help pinpoint the direction that the glacier once traveled. As rocks are carried by the ice and rubbed against each other, the sharp edges and corners are broken off into the meltwater. This polishing of the rocks makes rock flour, which can make the water appear cloudy.

The landscapes created by glaciers and areas of deposition are discussed in Chapter 7, "Deposition."

Waves

Waves are generated by winds in the atmosphere. The size and frequency of waves in the ocean are dependent on the length of open water that is available, which is called **fetch,** and the time that the waves have to build up in height. The more time and distance that the winds can blow, the larger the generated waves are. Wave heights are rarely more than 15 m under normal circumstances, but can reach 30 m in a hurricane.

Features of Waves

Figure 6-5 shows the measurement of wave height and wavelength.

Figure 6-5 Wavelength and height.

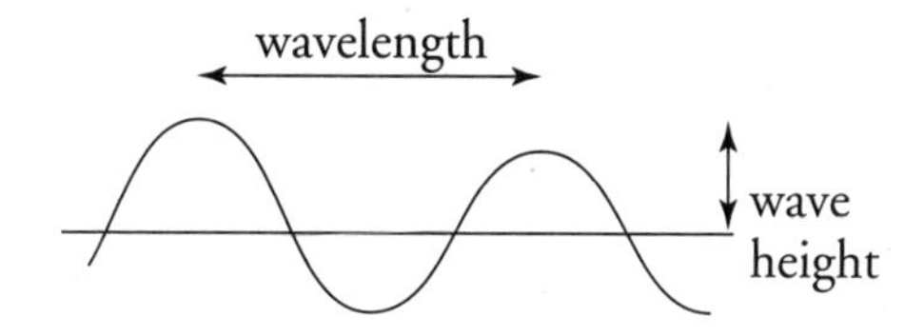

The **period** of a wave is how long the wave takes to move from crest to crest. As the waves approach the shore, the depth of the water decreases. The waves begin to break when the depth of the water equals one-half of the wavelength. Waves usually arrive in sets, with the waves in the middle being the largest. The pounding of the waves on the shoreline helps to weather the beach.

Tsunamis

Earthquakes generated underwater can create a **tsunami.** This is usually labeled as a "tidal wave," which is incorrect, since it has nothing to do with tidal variations in sea level. The size of the wave is dependent on the type of earthquake that occurred and the depth of the epicenter, and the shoreline.

Currents

As waves bring water to the beach, the water has to return to the ocean. This backwash is more commonly known as **undertow.** These are the dangerous currents that aren't easily seen and can pull tired or weak swimmers underwater. As the ocean water moves toward and away from the shore, another current develops. This **longshore current** moves parallel to the coast, transporting sand in a zigzagging motion with each wave You might have experienced this while playing in the ocean waves. As time goes on, you notice that you seem to be drifting along the shore and away from your blanket.

Another dangerous current that is created by the waves moving in and out from the shore are **rip currents.** These are very strong surface currents that move like small rivers perpendicular to the shoreline. They aren't usually more than 50 to 100 ft wide but can move at speeds up to 5 km/hr. If you are caught in a rip current, you should never try to swim back to shore against the current. You will get tired very quickly and that will cause you more problems. Instead, you should swim parallel to the beach. This will bring you out of the current and enable you to swim back to shore.

Erosion by Waves

Waves break down rock material through the energy and power of the wave and through abrasion. Many features can be created along the shoreline, depending on the type of rock that is present. Some of these features can include sea cliffs, sea caves, sea arches, and sea stacks.

Gravity

Gravity is the force that pulls on all objects. This includes the pull of rock fragments down slopes of hills and mountains.

Mass Movements

Slow movement of rocks and sediments is called **creep.** These movements are best observed with a time-lapse camera. If the movement is sudden, a landslide occurs. In the mountains, a landslide that occurs with snow, ice, and rock material falling is called an avalanche. When water saturated with clay and silt moves quickly, a mudflow or mudslide can result.

Chapter Checkout

Q&A

1. Which is the best example of physical weathering?
   - **a.** The cracking of rock caused by the freezing and thawing of water
   - **b.** The transportation of sediment in a stream
   - **c.** The reaction of limestone with acid rainwater
   - **d.** The formation of a sandbar along the side of a stream

2. The chief agent of erosion on Earth is

a. human beings.
b. running water.
c. wind.
d. glaciers.

3. Which agent of erosion forms U-shaped valleys?

a. Running water
b. Ocean currents
c. Wind
d. Glacial ice

4. Which statement best describes a stream with a steep gradient?

a. It flows slowly, producing a V-shaped valley.
b. It flows slowly, producing a U-shaped valley.
c. It flows rapidly, producing a V-shaped valley.
d. It flows rapidly, producing a U-shaped valley.

Answers: 1. a **2.** b **3.** d **4.** c

Chapter 7
DEPOSITION

Chapter Checkin

- ❑ Understanding the factors that affect the settling rate of particles
- ❑ Knowing the differences between deposition caused by running water, glaciers, and wind

Sediments that are transported by erosional forces are eventually dropped off. The manner of deposition and arrangement of the particles depends on the method of transportation. The deposition that occurs in streams differs from wind-driven particles and from sediments that were carried by glaciers.

Factors Affecting Deposition

The rate at which weathered particles are deposited can be affected by several factors. These can include the size, density, and shape of the particle. If you were to drop a handful of round quartz particles into a column of water, the largest ones would settle first. By changing only the density of the particles, different results would occur. If the particles were all the same size and shape, you would find that the more dense particles would settle first. The last factor is shape. Assuming that you had some particles that were made of the same material (same density and mass) but varied only in shape, the roundest particles would settle first. The flat particles would settle last. This type of particle sorting is called vertical sorting. The speed of the erosional agent is also factored in to determine the size of the particle that can be carried or deposited. Each of these types of erosional force is discussed in more detail later in this chapter.

Running Water and Wind

These two agents have one feature in common. They both rely on the speed of the running water or wind to carry particles. As the speed that the wind or water is moving decreases, the size of the particle that it is carrying gets smaller. As the water or wind slows down, the larger particles are deposited first. This helps to sort the particles by size.

The appearance of the sediments can tell how they were transported and deposited. Particles that were transported by running water and wind differ in the way that they look on the outside. Eroded fragments that were transported by running water have rounded edges. The sharp edges are worn away by the abrasion that takes place. Fragments that were affected by the wind have been sandblasted by smaller pieces of rock. This scratching of the surface creates a frosted appearance on the sediments.

Glaciers

As glaciers drag rock fragments along, the fragments are eventually deposited when the glacier stops. When the rocks are dragged along the ground, scratches in the surface, called **striations,** are created. They point in the direction that the glacier moved. The rocks that are moved along by the glacier are scratched and polished as well. The angular nature of the sediments stays intact. When the glacier stops advancing, the pile of rocks that it has been carrying gets dropped off. The pile of rocks left behind as the glacier melts is unsorted material called **till.** Smaller particles might be carried by the meltwater; this is called **drift.** The result of this drift creates **outwash plains** or sandy areas coming off of the moraine. These areas have sorted particles. **Erratics** are larger rocks that are deposited along the way. These are the big boulders that you might see in areas where they look out of place from the surrounding areas. These erratics are different types of rocks than the local bedrock or have a different size from the local rocks.

As the glacier moves downhill like a frozen bulldozer, **moraines** can develop, which are piles of rocks that are in a line. **Terminal moraines** are created as the glacier comes to a halt and leaves behind a pile of rocks in front of the glacier. Long Island, New York, was created by the last advance of the Laurentide glacier. It was actually two terminal moraines left behind. Rock piles that accumulate along the side edges of the glaciers form **lateral moraines.** In areas where two glaciers meet and combine forces, the lateral moraines merge and become a **medial moraine.** Ice blocks that are left behind after the glacier retreats form a small depression in the ground called a **kettle,** which can sometimes fill with water. These kettle ponds have steep sides.

Streams

As a stream gets into the later stages of its life, it develops areas of deposition in addition to the area at the end of the stream. The meandering loops of an old-age stream are systems of erosion and deposition. As the stream moves around a bend, the water on the inside of the bend slows down. An analogy for this would be a line of people trying to walk like a pinwheel. The people on the outside edge have a longer distance to travel. They have to walk faster to keep the line straight. The people on the inside walk a shorter distance, so they move at a slower pace. Larger particles carried by the stream are deposited because of the decreased velocity. While this happens, the water moving on the outside edge moves faster. The stream picks up larger particles, and the banks are eroded. This system of erosion/deposition perpetuates itself, causing the stream to meander even more. Eventually, some of the loops are cut off from the main flow, creating oxbow lakes.

Over time, the river may flood over its banks. The areas on the banks of the river are called **flood plains.** This is a natural occurrence that helps to replenish the land with nutrients.

As the stream empties into a larger body of water, such as a lake or ocean, the particles it is carrying are deposited. The stream slows down gradually, causing the particles to be sorted by size. The largest particles are deposited first, followed by smaller and smaller particles. These horizontally sorted particles will eventually be compressed into various sedimentary rocks, based on the grain size. In some areas, the deposition at the mouth of the river forms a triangle shape called a **delta.** The Mississippi River is a classic example of an old-age river showing meanders, flood plains, and a delta.

Ocean Waves

As waves crash onto the land, weathering, erosion, and deposition occur. If the waves are lower in energy, the beach is built up. The beach is the area between the high-tide and low-tide marks, made up of sand, rocks, and/or pebbles. During storms and other times of higher-energy waves, the beach particles are washed back into the ocean. The longshore current brings the sand down the coast to be deposited in another area. Several features are created by the movement of sand. A sandbar can form along the shore, just off the coast. If the sandbar becomes attached to the land, a spit is formed. Some sandbars are seasonal, and some become more permanent. A lagoon is the water between the sandbar and the mainland. Larger sandbars, which run parallel to the coast, are called barrier islands. Examples of these are Fire Island, New York; Atlantic City, New Jersey; Galveston, Texas; and Palm Beach, Florida. Severe storms can greatly affect these areas and the people who live there.

Wind

As with running water, the speed of wind determines the size of the particles that can be carried. The reverse is the case for the deposition of particles. As the speed of the wind decreases, the larger particles are deposited first. Sand dunes are created in areas where there is a lot of windblown material. The dunes are constantly moving and changing shape. The flow of air over the land causes the shape of the dune to be similar to an airplane wing. The backside of the dune is steeper and is the **slip face** (see Figure 7-1).

Figure 7-1 A dune's slip face.

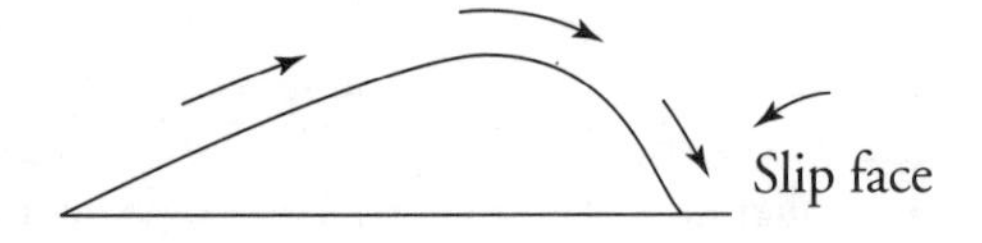

Mass Movements

Gravity is the main driving force for the mass movements of particles. The methods were described in Chapter 6, "Weathering and Erosion," but the results are generally the same. The particles are unsorted and angular. As gravity pulls the rock material down, it can pile up at the bottom of a cliff. This is known as **talus.** When small blocks of land are uplifted, rock material can move downhill, creating a **slump.**

Chapter Checkout

Q&A

1. Granite pebbles are found on the surface in a certain area where only sandstone bedrock is exposed. Which is the most likely explanation for the presence of these pebbles?
   - **a.** The granite pebbles were transported to the area from a different region.
   - **b.** Some of the sandstone has been changed into granite.
   - **c.** The granite pebbles were formed by weathering of the exposed sandstone bedrock.
   - **d.** Groundwater tends to form granite pebbles within layers of sandstone rock.
2. Unsorted piles of angular sediments were most likely transported and deposited by
   - **a.** wind.
   - **b.** glaciers.
   - **c.** ocean waves.
   - **d.** running water.
3. The particles in a sand dune are small and very well sorted, and have surface pits that give them a frosted appearance. This deposit most likely was transported by
   - **a.** ocean currents.
   - **b.** glacial ice.
   - **c.** gravity.
   - **d.** wind.
4. Compared to a low-density spherical particle, a high-density spherical particle of the same size will sink through water
   - **a.** more slowly.
   - **b.** more rapidly.
   - **c.** at the same rate.

Answers: 1. a **2.** b **3.** d **4.** b

Chapter 8
LANDSCAPE FORMATION

Chapter Checkin

- ❑ Understanding how weathering and erosion affect landscapes
- ❑ Learning about stream drainage patterns

The effect that weathering and erosion have on the surface of the Earth can shape how the land looks. The amount that surface bedrock is resistant to weathering can create some spectacular rock formations. The landscape of an area is determined by the hills, mountains, valleys, and streams of that region. The topography of the area is carved out by the climate, local bedrock, and influences made by humans. Uplifting and leveling forces are constantly at work on the land. Uplifting forces raise the bedrock upward creating mountains, plateaus. The forces of weathering and erosion wear down the land. These leveling forces move sediments back to water where new sedimentary rock is made.

Landscape Characteristics

The different agents of erosion can leave surface patterns that have specific characteristics.

Stream Patterns

The patterns seen in a stream system are dependent on the shape of the land. Smaller streams lead into larger streams. Eventually, they reach the mouth of the river, where the water empties into an ocean or lake. The region around the area of these streams is called the **drainage basin.** The outer edge of the drainage basin is the **drainage divide.**

Hillsides

The shape of a hillside can vary greatly depending on the climate and the type of rocks present. The basic parts of a hill are shown in Figure 8-1. The top edge is the waxing slope. This leads into the free face, followed at the bottom by the debris slope. Finally, the waning slope leads away from the hill.

Figure 8-1 The parts of a hill.

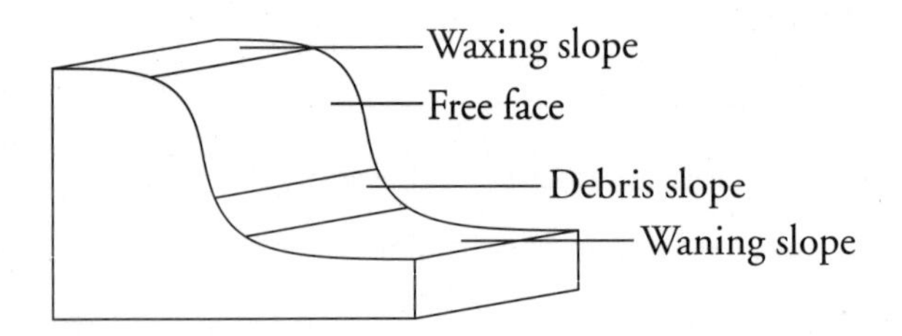

Landscape Regions

The main types of landscape regions are mountains, plateaus, and plains. The topographic relief of a specific area is the elevation difference between the lowest elevation and the highest point.

These regions are compared in Table 8-1.

Table 8-1 Landscape Regions

Region	*Elevation*	*Relief*	*Stream Speed*	*Origin of the Land*
Mountains	High	Large	Fast	Uplift
Plateaus	Moderate	Moderate	Medium	Small uplift
Plains	Low	Little	Slow, meandering	Sedimentation

Mountains

Mountains are the landscape region with the greatest relief. The bedrock in mountains can be igneous, metamorphic, or sedimentary. The sedimentary layers are folded or faulted. Mountains are generally created by converging plates, which are discussed in Chapter 13, "Mountain Building." Streams found here move quickly due to the high gradient. Examples of mountains are the Rockies and the Himalayas.

Plateaus

These areas have a moderate amount of relief over the region. The rolling hills associated with plateaus are found in areas of moderate elevation. This is due to the uplift of the entire region. The bedrock found here is generally of a sedimentary nature. Streams cut through, leaving wider valleys. The Appalachian and Colorado plateaus are good examples of these.

Plains

Plains have the least amount of relief. They are flat lands with low hills and are found in areas of low elevation. The underlying rock is usually sedimentary. The Gulf Coast and Atlantic Coast of the United States are good examples of plains.

Factors of Development

The rock layers present and the climate of the area help to determine what the surface features look like. The uplifting forces are the constructional forces. These increase the relief, elevation, and jagged look of an area. The leveling forces are working to erode the land back into the ocean. These smooth out the land and lower the relief and elevation.

Rock Types

Rocks such as metamorphic and igneous rocks take longer to erode and are more resistant to weathering than sedimentary rocks. Cliffs and escarpments are created when harder rocks are left behind as softer rocks are eroded away.

Climate

The climate of the area has an effect as well. In **arid** conditions, the surface is more angular and the hills are steeper. Humid climates, where more moisture is present, create landscapes that are more rounded. More vegetation exists in humid areas, keeping more of the soil from being washed away.

Human Interactions

People can have an effect on the landscape of a region. The areas carved out by construction projects, mining, the logging industry, and farming can drastically change the development of the landscape. Erosion caused by humans can be prevented. Farmers are implementing contour plowing. Using this method, farmers plow and plant the land along flat surfaces, like the way contour lines run, instead of up and down hills. In areas that have been strip-mined, the removed topsoil can be replaced. Vegetative coverings are planted to help keep the soil from being eroded.

Drainage Patterns

The pattern of the stream as seen from above can tell you about the topography of the land. Figure 8-2 shows each of the major drainage patterns. A dendritic pattern is characteristic of flatter land with uniform bedrock. If the water drains outward from a volcano or a dome mountain, a radial pattern develops. Rectangular-shaped drainage streams form from tilted, faulted, or folded rock layers. A dome mountain with upturned layers will have an annular pattern.

Figure 8-2 Drainage patterns.

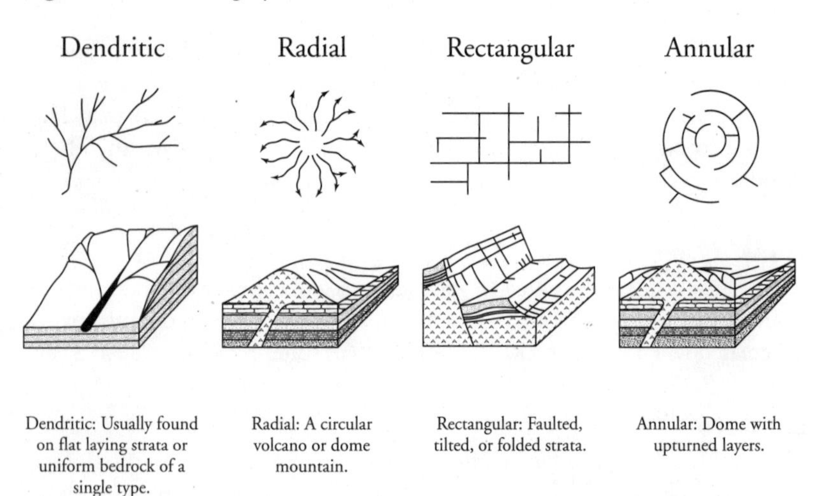

Features

Landscape regions are areas that have similar landscape characteristics. These include existing bedrock, elevation, topographic relief, streams, drainage patterns, and soil. The boundaries between the landscape regions are usually defined. The bedrock might be different; fault zones might exist. These are a result of the forces that created the region. Cliffs, ridges, or mountain edges can be found in this boundary area.

Chapter Checkout

Q&A

1. Landscape regions in which leveling forces are dominant over uplifting forces are often characterized by
   - **a.** volcanoes.
   - **b.** mountain building.
   - **c.** low elevations and gentle slopes.
   - **d.** high elevations and steep slopes.
2. A landscape with only intermittent (not permanent) streams and no drainage connected to the ocean would most likely have what type of climate?
   - **a.** Arid
   - **b.** Humid
   - **c.** Hot and wet
   - **d.** Cool and wet
3. Which change would occur in a landscape region where uplifting forces are dominant over leveling forces?
   - **a.** Topographic features become smoother with time.
   - **b.** A state of dynamic equilibrium will occur over time.
   - **c.** Streams decrease in velocity with time.
   - **d.** Hill slopes increase in steepness with time.
4. As human population increases, the amount of alteration of the landscape by human activity will most likely
   - **a.** increase.
   - **b.** decrease.
   - **c.** remain the same.

Answers: 1. c **2.** a **3.** d **4.** a

Chapter 9

WATER IN THE GROUND

Chapter Checkin

- ❑ Knowing about the amount of useable water on Earth
- ❑ Determining how precipitation can infiltrate the ground
- ❑ Understanding how water wells work

When water reaches the surface of the Earth, it has several different pathways it can take as part of the water cycle. It can evaporate into the air, run off into a stream, or sink into the ground. Transpiration by plants puts water back into the atmosphere, where it will later condense into the liquid or solid form. This chapter deals with the factors that affect how water sinks into the ground and its effect underground.

Water in the World

The water supply of the Earth is finite but is constantly being used and recycled. The actual amount of water that is useable is very small. About 97.2 percent is salt water and 2.8 percent is fresh water. Of that 2.8 percent, only about 0.6 percent is drinkable. Figure 9-1 shows how the water supply is divided up.

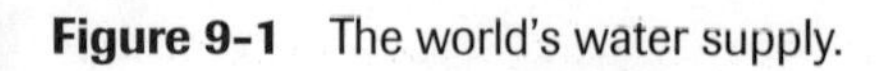

Figure 9-1 The world's water supply.

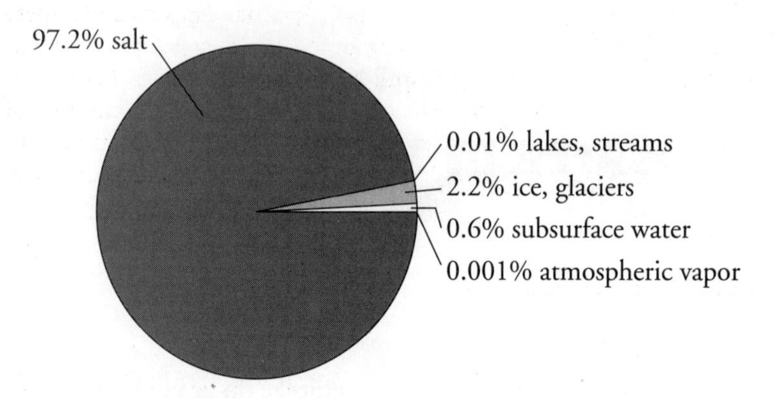

Water Budget

The water budget describes the amount of incoming precipitation and evaporation for an area. The amount of precipitation varies greatly depending on where you are in the world. The amount of evaporation for a region is dependent on the concentration of sunlight for that area and the humidity in the air.

One way to think about the water budget is to look at a glass of water. Start off with a full glass. Over time, water evaporates from the glass, or you take a drink. This **usage** is similar to water being used in the soil by plants or people and by evaporation from the sun. This is also a time when the amount of water being poured into the glass is less than the amount being taken out. This mainly happens during the later spring and early summer.

When the cup is empty and any water being put into it is used immediately, a **deficit** occurs. This is the time of drought for an area, which happens often in August. There is not enough water to meet the needs of an area. If the cup stays empty and is not refilled, a prolonged period of drought is experienced by that area. When the amount of water in the glass starts to fill back up, this is a time of **recharge.** During this time of recharge, the amount of water that is being put in is greater than the amount being taken out. The fall season is a common time for this occurrence. Eventually, the cup is filled up.

If more water is poured into the cup while it is full, a **surplus** condition results and runoff occurs. Flooding is a problem that happens under these conditions. This is common in the springtime when water from the melting snow combines with spring storms and saturated ground. These conditions can be described through graphs. They show the amount of precipitation that an area receives and the amount of evaporation and transpiration that occurs. Figure 9-2 shows sample water budgets for representative areas.

Figure 9-2 Sample water budgets.

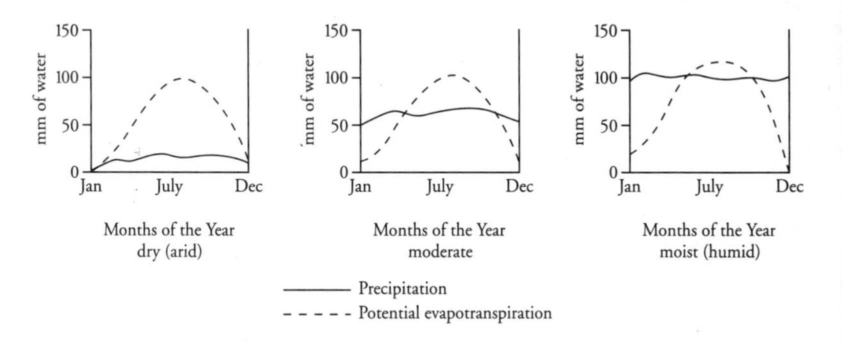

Water in the Ground

The ability of rocks and the ground to hold water is dependent on several factors.

Ability of Rocks to Hold Water

The amount of open pore space available between the particles is known as **porosity.** This space can vary greatly depending on the shape of the particle. An area with round particles has a lot more space available than an area with angular particles. The angular pieces fit together more closely and fill in the pore spaces better. Another variable is the sorting of the particles. Poorly sorted grains have less space available as well. The spaces between the larger particles are taken up by the smaller pieces. In areas where the grains are well sorted (lots of similar-sized particles), more pore space is available to hold water.

Transmission of Water

Any precipitation that sinks into the ground is known as **infiltration.** The rate at which this water penetrates into the ground is the **permeability** rate. If water cannot pass through the rock, as is the case with clays and shales, the rock is considered to be **impermeable.** Once in the rock and soils, water can stick to the particles. This is **capillary** water, and it can actually work its way up toward the surface. This movement against the force of gravity is due to the forces of adhesion to the rock particles and cohesion with water molecules

Water Table and Wells

As rain sinks into the ground, it may reach an impermeable layer. The underground pore spaces start to fill up with water. The area filled with water is the **zone of saturation.** The surface of this zone is the **water table.** Above the water table is the **zone of aeration.** This zone is where the pore spaces are filled with a mixture of air and water. The height of the water table varies for every area and is seasonal within one area. If the water table is high enough, it can leak out onto the surface of the Earth. A spring is created. By drilling down into the ground, water can be pumped up to the surface to be used by people, which is called a well. If the zone of saturation is located between two impermeable layers, an artesian well can be made. These wells don't need a pump to bring the water to the surface but are usually deep in the ground. The water is forced upward, trying to balance itself. The water wants to be at the same height as it is on the other end of the zone of saturation. Figure 9-3 shows this.

Figure 9-3 Artesian and pump wells.

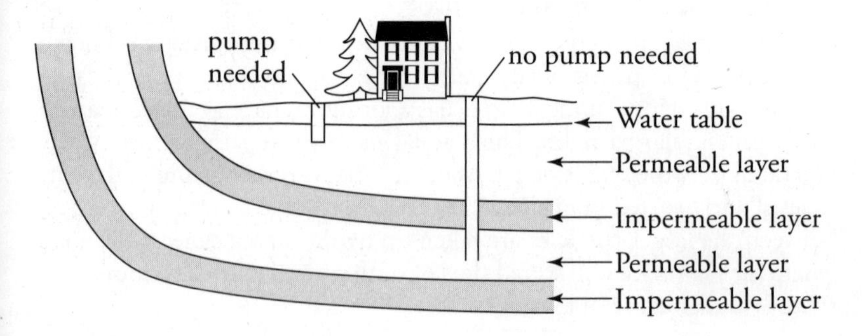

Groundwater Characteristics

In general, the water in the ground is cool in temperature. If the water is heated by the Earth, hot springs or geysers could form. The ground acts as a filter, cleaning the water so that people can drink it. The minerals that are in the soil and rocks can get into the groundwater. In some regions, this can mean that sulfur or iron is in the drinking water. The pH of the water can be affected by the soils as well. Remember that the amount of water that is actually usable is very small—about half of 1 percent. We need to conserve the water that we have and prevent the pollution of future water supplies.

Conservation

The amount of actual water that is drinkable on Earth is very small in comparison to the amount of water available. The prevention of future pollution is extremely important. Equally important is to clean up the polluted drinking water that we already have. As the world's population increases, so will the need for more water. This idea of conservation and cleanup also translates to our air. Since the 1970s a great deal of awareness has been made to aid this effort. A lot of areas have been cleaned up, but there is plenty more to be done. Although humans have been on the Earth for a short time (geologically speaking), we have done a lot of damage to the Earth. Most of this has occurred in the last few hundred years. We should be just as involved in its cleanup. Hopefully we can accomplish this in less time than it took us to create the problems.

Caves

Water sinking into the ground can dissolve limestone bedrock. Caves and caverns can be created. If the layers above the caves can't support the weight of the land above, a sinkhole is formed as the rock layers collapse. As the water drips down from the ceiling of a cave, it might evaporate before it reaches the floor. The minerals are left behind, forming a "rock icicle." These are **stalactites.** Some of the mineral-laden water reaches the floor. **Stalagmites** are formed upward as a result of this. Stalactites are V-shaped (they come to a "tight" point), and stalagmites are wider (mightier). Stalactites hold "tight" to the ceiling, whereas you "might" trip over a stalagmite. If they eventually meet, a **column** is formed.

Chapter Checkout

Q&A

1. Which condition is most likely to cause surface runoff during a rainstorm?
   - **a.** The permeability of the soil is greater than the rate of rainfall.
   - **b.** The porosity of the soil is greater than the amount of rainfall.
   - **c.** The surface slope allows for maximum infiltration.
   - **d.** The surface soil is saturated.

2. A rock with high porosity will probably
   - **a.** be resistant to weathering.
   - **b.** be composed of large grains.
   - **c.** have a large percentage of space between the particles.
   - **d.** have a small percentage of rounded particles.

3. Stream discharge normally would be highest during a period of
   - **a.** recharge.
   - **b.** deficit.
   - **c.** usage.
   - **d.** surplus.

4. Water budget graphs for four cities, A, B, C, and D are shown in Figure 9-4.

Figure 9-4 Water budgets for four cities.

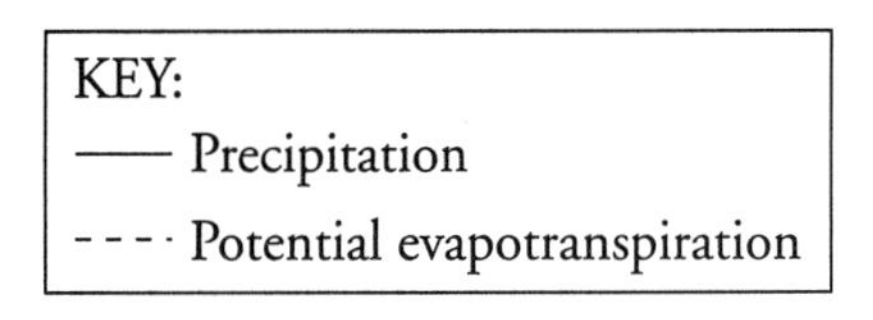

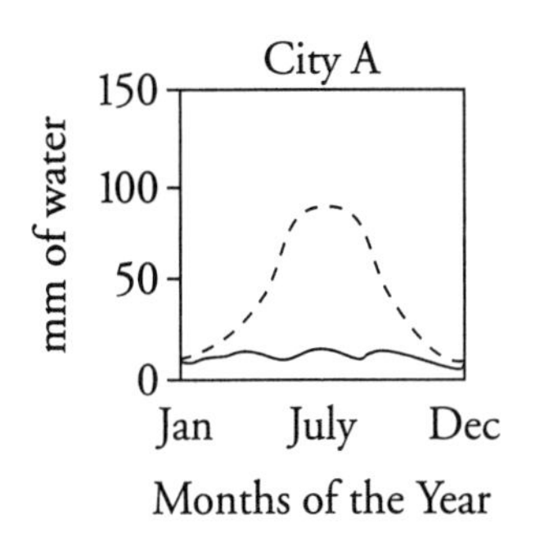

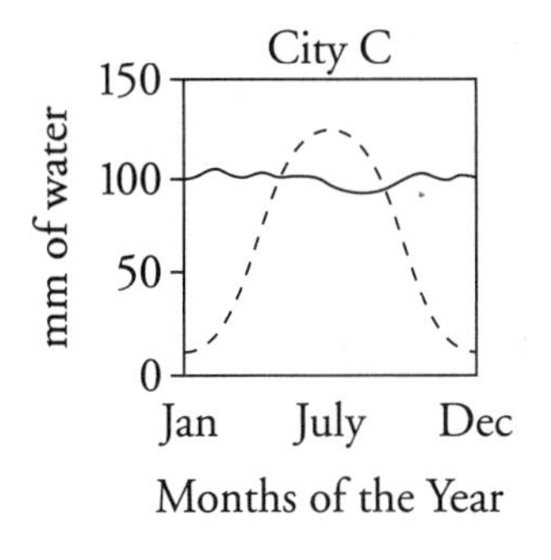

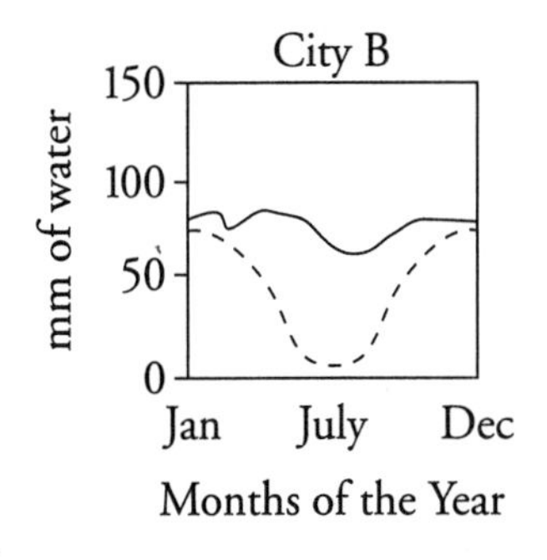

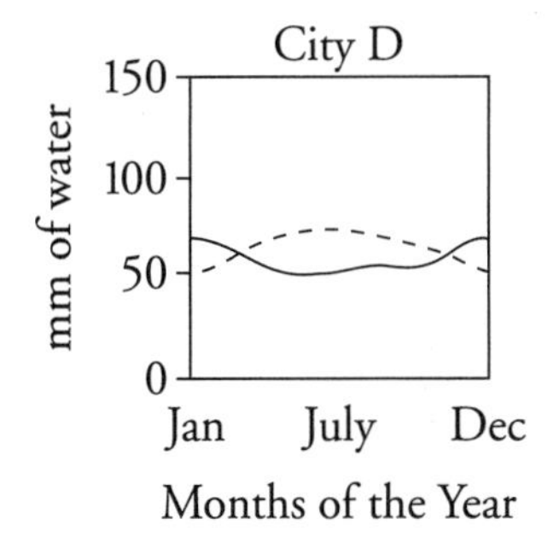

Which city would have the driest climate?

a. A
b. B
c. C
d. D

Answers: 1. d **2.** c **3.** d **4.** a

Chapter 10

PLATE TECTONICS

Chapter Checkin

- ❑ Knowing the Earth is made of layers and is covered by moving plates
- ❑ Understanding the forces that cause plate tectonics

The Earth is covered by rock plates that float on the surface of the Earth. As they move, different events can happen. Earthquakes, volcanic activity, and mountain building can occur as a result of these events.

Layers of the Earth

The Earth can be divided into four distinct layers. Each layer has specific characteristics that separate it from the other layers. The surface of the Earth is covered by 12 major plates and a few other smaller plates. These plates are part of the **lithosphere** and are about 100 km thick. The oceanic plates are made from basalt, whereas the continental plates are made from granite, which is less dense than basalt. The difference in density between the plates will have an effect on the results of plates converging.

The outermost layer of the Earth is the **crust.** The thickness of the crust varies from 10 km thick below the oceans to 65 km thick under the continents. Even though the crust is relatively thin, man hasn't gone through it yet. The next layer down is the **mantle,** which is about 2,900 km thick and is rich in iron, magnesium, and silicon-containing rocks. The mantle can be subdivided into two layers. The outer layer is the **asthenosphere,** which is a plastic-like layer. The convection currents found here cause the plates to move. The **outer core** is about 2,250 km thick and is the next layer in toward the center of the Earth. This layer is made of liquid iron and nickel and surrounds the **inner core.** The inner core is solid iron and nickel. Figure 10-1 shows the layers of the Earth.

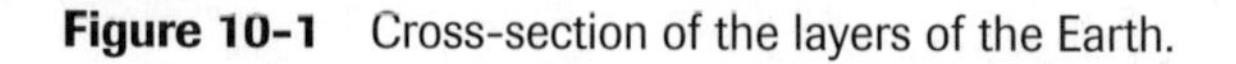

Figure 10-1 Cross-section of the layers of the Earth.

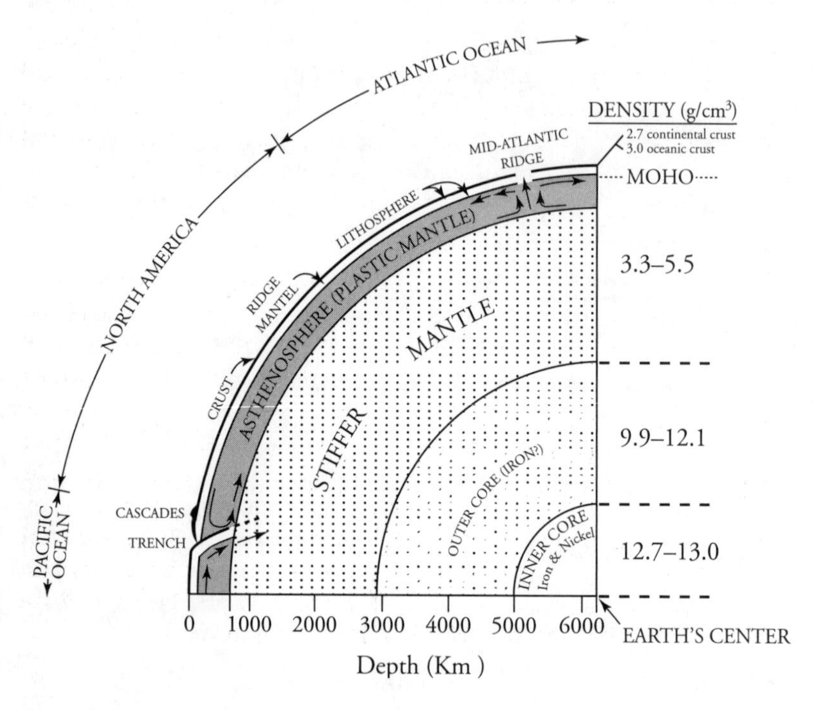

Evidence of Crustal Movement

In the early twentieth century, maps of the world were improving in their accuracy. Although the continents looked like they would fit together like puzzle pieces, there wasn't any evidence until Alfred Wegener proposed his theory of Continental Drift in 1912. He found Mesosaurus fossils in Brazil and South Africa. Both sets of fossils dated back about 270 million years ago. These were the only places that these fossils were found. Wegener also noted that there were similar rocks and mineral deposits found on both of these continents in corresponding locations. More scientific support for the theory wouldn't come until the 1960s.

The correlation between earthquakes and volcanoes is very high. They generally occur along the boundaries of plates. The Ring of Fire in the Pacific Ocean accounts for 90 percent of the earthquakes in the world. The discovery of fossils from tropical plants and animals in colder regions helped to support the idea that the continents are moving, in addition to glacial evidence in now-tropical areas.

Plate Tectonic Theory

The idea of the continents fitting together seems simple and logical. According to computer models, at one time all of the continents were together, forming a landmass called Pangaea. This supercontinent broke apart about 230 million years ago, forming Laurasia to the north and Gondwanaland to the south. A look at the coastlines shows that the continents seem to fit together with few gaps. A closer look at the continental shelves shows they fit together even better.

The coasts of the continents are constantly being changed by waves, rivers, and precipitation. The driving force behind the movement of the plates, however, lies further underground. The inner layers of the Earth heat up rock in the asthenosphere, which rises upward due to a lower density. As the rock reaches the outer edge of the mantle, it flows outward. This movement pulls the plates along until they meet another plate. The process for this movement is similar to the motion of the wheels on a bulldozer. The plates would sit on top of the tracks and move. Eventually the rocks cool, increase in density, and sink back into the mantle. Figure 10-1 shows that the Mid-Atlantic Ridge marks the location where the rock moves upward. This pathway continues until plates meet at the Pacific Ocean and are drawn back into the Earth, where the materials are recycled.

Plate Boundaries

Three types of plate boundaries exist: **diverging, sliding** or **transform,** and **converging.** These are the interfaces between plates, which are created by the relative movement of the plates to each other. Figure 10-2 shows the major plates and their direction of movement.

Figure 10-2 Tectonic Plates and their movement.

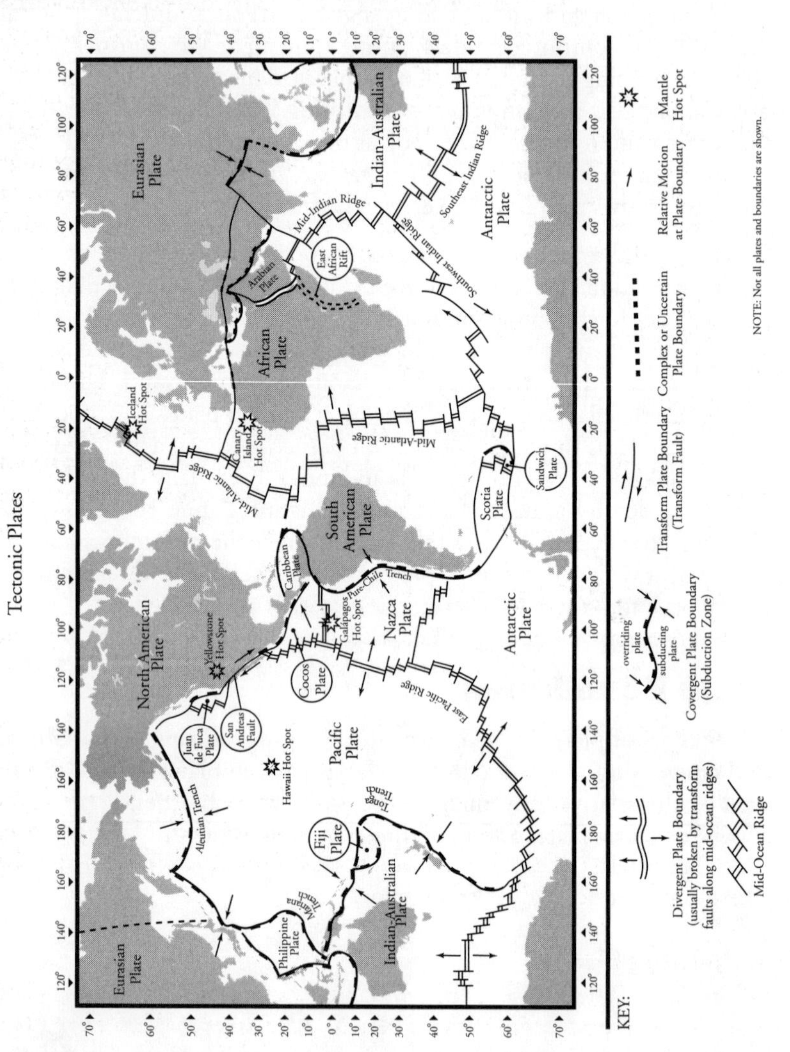

Diverging Boundaries

Divergent plate boundaries are areas where plates are moving away from each other. Magma from below rises up and forms a **mid-ocean ridge.** As the plates pull apart, a space between them called a **rift valley** is created. Oceanographic trips to these areas have uncovered entire ecosystems. Hot springs that arise here help the chemosynthetic organisms thrive. They use sulfur instead of sunlight to create energy. Many new species have been discovered in these geothermal vent areas, including giant clams, large tube worms, and blind crabs! Examples of diverging boundaries are found in the Mid-Atlantic Ridge and the East Pacific Rise. The East African Rift valley is a divergent boundary above sea level.

Sliding Boundaries

When the plates slide past each other, a sliding boundary exists. This is also called a **transform boundary.** Due to the increased friction, these areas have many faults. Faults are breaks or cracks in the crust along which movement has occurred. The San Andreas Fault in California moves about 5 cm per year. This is part of the larger sliding boundary between the Pacific Plate and the North American Plate.

Converging Boundaries

As two plates come together, several scenarios can result. These convergent boundary zones fall into two categories: collision and subduction.

Collision between Plates

If the two plates that crash into each other are continental plates, they both crumple upward, forming a single land mass. The boundary between India and China exemplifies this. India is moving north at about 5 cm per year, which is creating the Himalayan Mountains. This mountain range is the tallest in the world, and is still growing.

Subducting Plates

When a continental plate collides with an oceanic plate, the oceanic plate plunges downward while the continental plate stays at the surface. The oceanic plate is pulled under due to the density differences of the plates. Ocean plates are made of basalt, which has a density of about 3 g/cm^3, whereas the density of the granite of the continental plates is about 2.7 g/cm^3. The oceanic plate is considered to be the **subducting** plate, whereas the continental plate is called the **overriding** plate. This can also occur as the result of a collision between two oceanic plates. A deep sea **trench** is formed just off the coast of the land. The Mariana Trench off the coast of Japan is 11 km deep and is the deepest location on Earth. The pressure at this depth is about 100 times the pressure at sea level. As the subducting plate is pulled deeper into the Earth, it heats up and expands. The rocks of the overriding plate are forced upward, forming mountains. A mountain chain is formed on the continent, running parallel to the boundary. The Andes Mountains in South America were formed this way when the Nazca Plate collided with the South America Plate. If the pressure and temperature of the subducting plate reach the melting point, magma is formed. This magma can rise up to the surface to create a volcano.

Figure 10-3 compares the different plate boundaries.

Figure 10-3 Types of Plate boundaries.

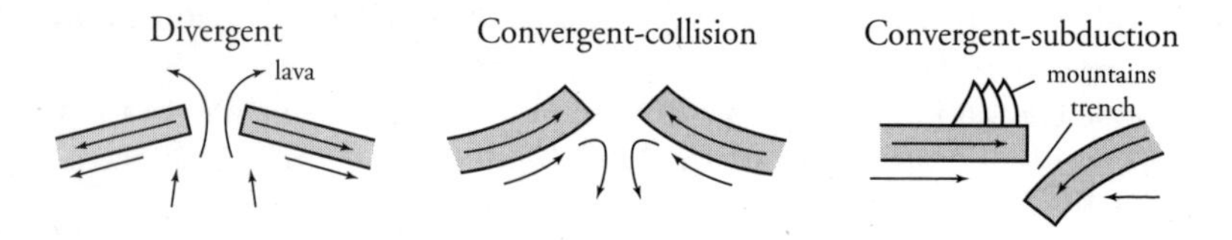

Hot Spots

Weak or thin areas in plates can allow for magma to rise up and reach the surface. These **hot spots** are areas where new lands are formed. The Hawaiian Island chain was formed over a hot spot. Over time, the Pacific Plate has moved. As it moves, another island is formed. The newly formed islands move away from the hot spot, where they are exposed to weathering and erosion by the ocean waves. A closer look at this area reveals that the island chain actually continues underwater along the Emperor Mountain chain. The Canary Islands, Iceland, and Yellowstone Park in the United States are other active hot spots. Figure 10-2 shows the locations of these hot spots.

Pole Reversal Patterns

Over time the magnetic poles of the Earth have switched places. The reason for this is not known, but each swap has taken place over a time span about 1,000 years. This is a short time, geologically speaking. The flip-flop of the magnetic field has been recorded in the rock record. There have been four major reversals of the magnetic fields in the last 4 million years. As plates diverge, lava comes out. The magnetic particles that are present in the liquid rock align themselves with the magnetic north and south poles. The rock, which is underwater, cools quickly, "freezing" these particles in place. The youngest rocks are found along the ridge between the plates. As you move away from the divergent boundary, the age of the rocks increases. Figure 10-4 shows the pattern of reversal as recorded by the rocks.

Figure 10-4 Patterns of pole reversal.

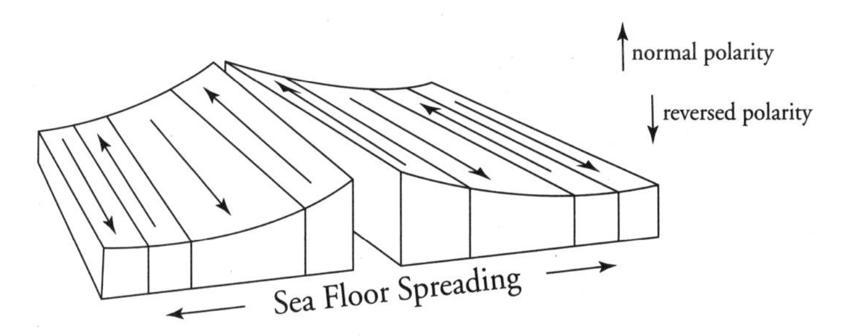

Chapter Checkout

Q&A

1. Which statement best supports the theory that all the continents were once a single landmass?
   - **a.** Rocks of the ocean ridges are older than those of the adjacent sea floor.
   - **b.** Rock and fossil correlation can be made where the continents appear to fit together.
   - **c.** Marine fossils can be found at high elevations above sea level on all continents.
   - **d.** Great thicknesses of shallow-water sediments are found at interior locations on some continents.

2. Igneous materials along oceanic ridges contain magnetic iron particles that show reversal of magnetic orientation. This is evidence that
   - **a.** volcanic activity has occurred constantly throughout history.
   - **b.** the Earth's magnetic poles have exchanged positions.
   - **c.** igneous materials are always formed beneath oceans.
   - **d.** the Earth's crust does not move.

3. The thinnest section of the Earth's crust is found beneath

a. oceans.
b. coastal plains.
c. mountain regions.
d. desert regions.

4. The rock found between 2,900 km and 5,200 km below the Earth's surface is inferred to be

a. an iron-rich liquid.
b. an iron-rich solid.
c. a silicate-rich liquid.
d. a silicate-rich solid.

Answers: 1. b **2.** b **3.** a **4.** a

Chapter 11
EARTHQUAKES

Chapter Checkin

- ❑ Understanding the causes of earthquakes
- ❑ Determining the epicenter and time of an earthquake

The surface of the Earth is in constant motion. As rocks break, a tremendous amount of energy is released, which creates waves that become earthquakes. The damage done and the effect on people depends on the strength of the earthquake and its location.

Areas of Frequent Crustal Movement

Earthquakes occur every 30 seconds. This adds up to more than 1 million each year, yet many of them are barely felt. About 3,000 of these earthquakes actually move the plates, with only several hundred quakes significantly moving the plates. The most severe changes and movements occur during larger earthquakes, which number about 20 per year. Most of the earthquakes occur along plate boundaries. The correlation between the locations of earthquakes and volcanoes is high.

Earthquake Waves

Earthquake waves are generated when rocks that have been bent to such a great extent break. The stress that is released results in an enormous amount of energy. This is the elastic-rebound theory. An analogy for this theory, on a very small scale, is that if you take a pencil and bend it at the middle, eventually it will break. But if you look carefully, you will notice that the broken ends of the pencil waver back and forth immediately after the pencil snaps in two. This is what generates the earthquake waves.

The depth of the earthquake depends on the type of boundary that causes the quake. Earthquakes originating along the San Andreas Fault are less than 30 km deep, whereas quakes starting along a subduction zone can be 700 km deep. The actual location of the earthquake is called the **focus.** The location on the surface directly above the focus is called the **epicenter.** This is the closest that we can get to the focus. An earthquake with a shallow focus causes damage to small areas, but it is more intense. As the depth of the focus increases, so does the damage area. If the earthquake is small with a deep focus, then the energy is dispersed over a larger area creating only a small tremor at the surface. If the ground is sandy or soft, then liquefaction can occur. The ground behaves as if it was a liquid. This can cause a lot of damage to structures, gas lines, and electric lines.

There are three different types of waves that are generated by an earthquake: **P-waves, S-waves,** and **L-waves.** These three types are discussed in the following sections.

P-Waves

The fastest wave that is caused by an earthquake is a primary wave. A **P-wave** moves twice as fast as an S-wave. It is a compressional wave, which travels through solids, liquids, and gases. The wave moves objects back and forth.

S-Waves

Although **S-waves** are slower than P-waves, they do more damage. They are shear waves that move up and down at right angles to the direction of motion. The motion is similar to a snake moving. S-waves travel only through solids. They don't move through liquids or gases. Remember to associate S-waves with the words *secondary, slower, shear, solids,* and *stronger.*

L-Waves

As the primary and secondary waves reach the surface, they combine to form a third type of wave. This is a longitudinal wave, or **L-wave.** This is the slowest wave and moves outward at speeds of about 3 km per second, like ripples on a pond.

Shadow Zone

The waves generated by an earthquake move in all directions. As the waves change layers in the Earth, the speed changes. There is a drastic change in their velocities at the interface between the crust and the mantle. The discontinuity here is the Moho, named for Andrija Mohorovicic, a Yugoslavian scientist who studied seismograms of minor earthquakes. As S-waves hit the outer core of the Earth, they are absorbed by the liquid rock. P-waves pass through, but are refracted. The result of this is an area of the other side of the Earth that doesn't receive waves initially. This is the **shadow zone.** Some waves that are reflected off of the Earth may reach these points, but with a lot less energy.

The layers of the Earth have been determined by earthquake data. By recording the location and strength of many earthquakes, the model of the Earth has been created.

Epicenter Location

Many earthquakes that occur are never felt or seen by humans, but a very sensitive instrument called a **seismograph** can detect them. **Seismograms** are the records made by these machines. Horizontal and vertical motions are recorded. Newer devices using laser and satellite tracking devices are being implemented as the technology advances. The seismogram records the waves as they pass through the station. The time difference between the arrival of the P- and S-waves is noted. By using the chart, as seen in Figure 11-1, the distance to the epicenter can be determined. By plotting this distance on a map, a locus of points making a circle is formed. The epicenter of the earthquake could be anywhere along the circle. The addition of the data from two more stations can pinpoint the epicenter. Three circles are drawn, one from each station, and where they intersect is the epicenter of the earthquake. The depth of the earthquake's focus can be determined by the lag time of the L-waves.

Figure 11-1 Earthquake P-wave and S-wave travel time graph.

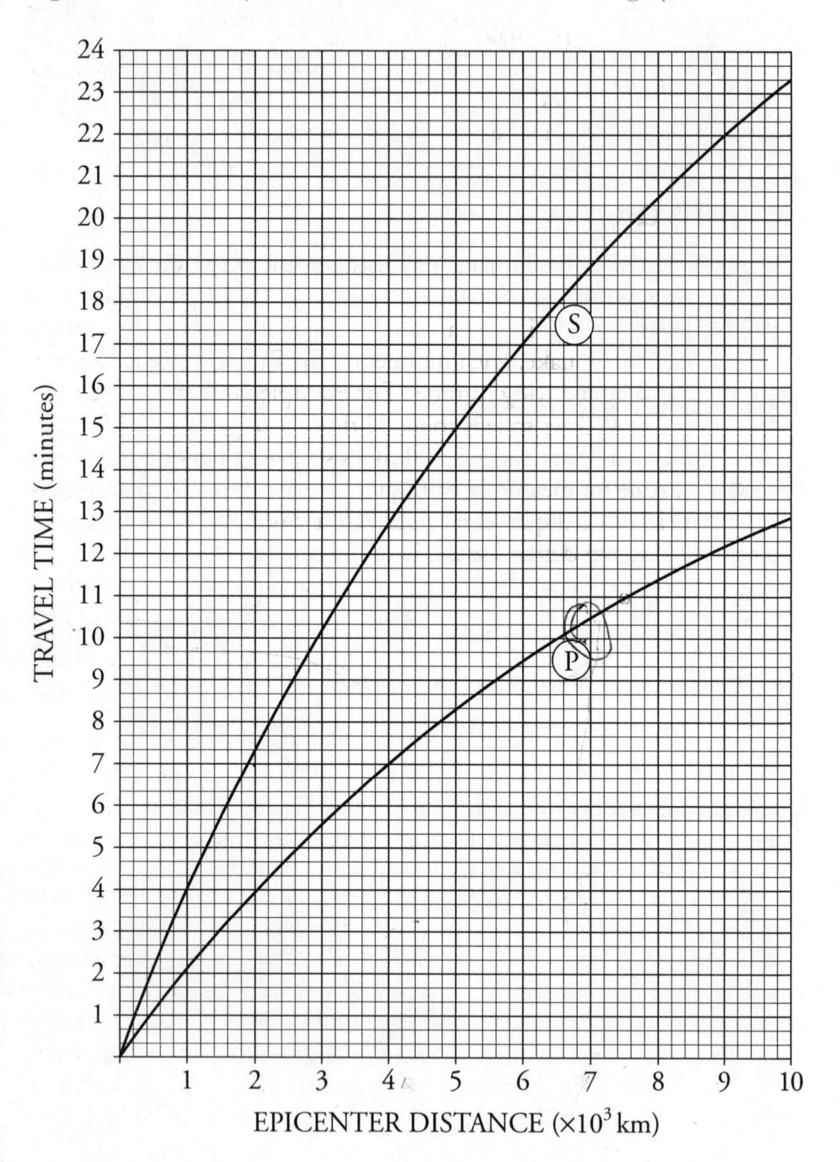

Origin Time

The time that the earthquake occurred can also be determined by in Figure 11-1. If the distance to the epicenter is known, the chart will tell you how long the P- and S-waves have been traveling. You can use this information to determine the time that the earthquake occurred.

Magnitude

The magnitude of an earthquake is a measurement of the size or amount of energy that was released. The **Richter scale** was developed in the 1940s by Charles F. Richter. It is a logarithmic scale that measures the energy released by the earthquake. Each number on the Richter scale is 32 times stronger than the preceding number. For example, an earthquake measuring 4.0 on the Richter scale is about 32 times stronger than a 3.0 earthquake. Similarly, a 7.0 quake is 1,000 times stronger than one measuring 5.0. Another scale for measuring earthquakes is the **Mercalli scale.** It measures intensities of earthquakes on a scale from I to XII and uses different damage to objects to determine the intensity scale

The strongest earthquake ever recorded was a 9.5 on the Richter scale in Chile in 1960. A series of notable quakes hit the town of New Madrid, Missouri, in 1811 and 1812. Three earthquakes with magnitudes of 8.6, 8.4, and 8.7 were so intense that they traveled through the solid rock of the United States and rang church bells in Boston, Massachusetts.

Tsunami Formation

Earthquakes that originate under the oceans can generate ocean waves. These waves reach the shorelines some time later, depending on the distance to the epicenter. The size of the tsunami waves is difficult to predict because the waves don't appear until they reach shallow water. Normal and reverse faults generally create larger waves than strike-slip faults. The steady flow of water and waves can cause tremendous damage to a low-lying area. One of the largest earthquakes of the 20th century was in Alaska in 1964. This 9.2-magnitude quake lasted more than five minutes, had land waves of 3 ft high, and created a tsunami that was about 21 m high. See Chapter 13, "Mountain Building," for more information on faults.

Predicting and Preparing for Earthquakes

The accurate prediction of earthquakes is an extremely difficult task. There are no accurate methods in place at this time. Some ideas being investigated are the detection of pre-earthquake tremors and the presence of radon in well water. As mentioned earlier in this chapter, satellites and lasers are being used to detect the slightest of movements in the Earth's crust.

The initial problems that arise from earthquakes are the collapse of buildings and explosions and fires, caused by the severing of electrical wires and gas lines. Longer-term problems are the spread of disease from broken sewage lines, water shortages, food shortages, and tsunamis. Some of these problems can be prepared for. By gathering an emergency kit, you can be better prepared for an earthquake. The kit should include at least: working flashlights, canned food, manual can opener, first aid supplies, battery operated radio and tissues. Water for drinking should be available as well. Other preparations must be addressed by engineers, such as building structures that can withstand earthquake waves.

Chapter Checkout

Q&A

1. A seismic station is 2,000 km from an earthquake epicenter. How long does it take for an S-wave to travel from the epicenter to the station?

a. 7 minutes, 20 seconds
b. 5 minutes, 10 seconds
c. 3 minutes, 20 seconds
d. 4 minutes, 10 seconds

2. Through which zones of the Earth do primary waves (P-waves) travel?

a. Only the crust and mantle
b. Only the mantle and outer core
c. Only the outer core and inner core
d. The crust, mantle, outer core, and inner core

3. A seismic recording station is 7,700 km from the epicenter of an earthquake. If the P-wave arrived at 2:15 P.M., at approximately what time did the earthquake occur?

a. 1:56 P.M.
b. 2:00 P.M.
c. 2:04 P.M.
d. 2:08 P.M.

4. Tsunamis can be directly caused by

a. offshore surface ocean currents.
b. gravitational effects of the moon.
c. underwater earthquakes.
d. underwater lava flows at mid-ocean ridges.

Answers: 1. a **2.** d **3.** c **4.** c

Chapter 12
VOLCANOES

Chapter Checkin

- ❑ Knowing where volcanoes are located
- ❑ Understanding how volcanoes form
- ❑ Explaining the structure of a volcano

There are about 500 active volcanoes on the Earth, with about half of these located around the Ring of Fire in the Pacific Ocean. About 50 volcanoes are located in the United States, which ranks third worldwide. The countries with the most volcanoes are Indonesia and Japan. **Active volcanoes** have erupted in recorded history. **Dormant volcanoes** have not erupted during this time.

Location

If you were to plot the location of volcanoes and earthquake epicenters, you would find a high correlation between the two. Most volcanoes are located near plate boundaries. The Ring of Fire is the boundary surrounding the Pacific Ocean. It is an area of many active and inactive volcanoes. Other regions of frequent volcanic activity are the Mediterranean, Africa, and Asia Minor.

Formation

Deep within the Earth, the temperature can reach over 6,000°C. The cause of this heat is three main sources. The radioactive decay of several elements, such as uranium and potassium, gives off energy. Some of the heat from when the Earth was first formed was trapped in the inner layers. The last source is the heat generated by the friction of plates rubbing against each other. Underground rock that becomes molten due to this heat becomes less dense and rises toward the surface. This magma melts other rocks with which it comes in contact. As it reaches the surface, it exits through a crack or thin area of the Earth's crust.

Magmas that are felsic have a higher silica content, are lighter in color, and are thicker. These felsic magmas therefore move slowly. Mafic magmas are low in silica, darker in color, and thinner. These flow more easily than felsic magma. The magma interacts with the air and chemical changes take place. The new substance is lava. The lava flows outward and downhill, and then cools. The speed of the lava depends on its composition. Mafic lava flows faster than felsic lava. Faster lava flows allow for gases to escape more easily. Thicker lava holds more gases and produces more explosive eruptions. The new rocks formed when the lava and magma cool are igneous rocks.

As more lava is forced out of the Earth, layers of rock and ash form a cone. Inside the magma there are dissolved gases of water vapor, carbon dioxide, and sulfur. The intense heat and pressure of the gases can make the magma explode out of the Earth with tremendous force.

Structure

There are three main types of volcanoes. The basic structure of these volcanoes is similar, which is shown in Figure 12-1. Volcanoes can erupt in several ways, with a variety of materials that erupt out from them. Each type of volcano has characteristic matter that it spews out. Any magma that cools and hardens before reaching the surface is called an **intrusion.**

Figure 12-1 Volcanic structure.

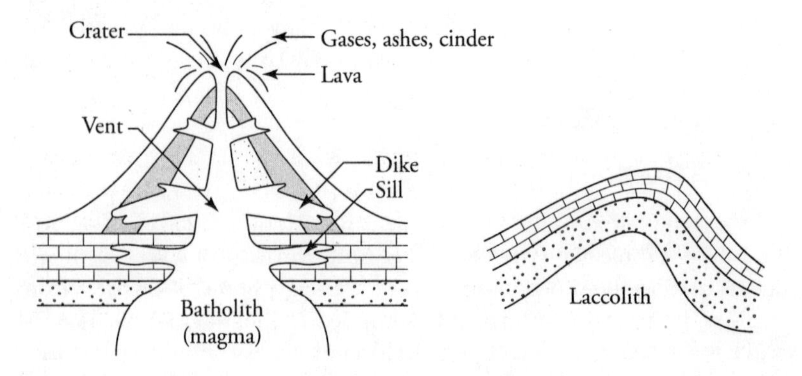

Cones

The structure of the volcano can take different shapes depending on how it is formed. A **shield cone** is a broad cone resulting from smooth lava flows. The layers build upon each other gradually. A **cinder cone** forms from ash and loose rock that is spewed out explosively from the volcano. Cinder-cone volcanoes are steeper than shield volcanoes. Figure 12-2 shows the two cone types. A **composite volcano** consists of alternating layers of solidified lava and rock particles.

Figure 12-2 Volcanic cone types.

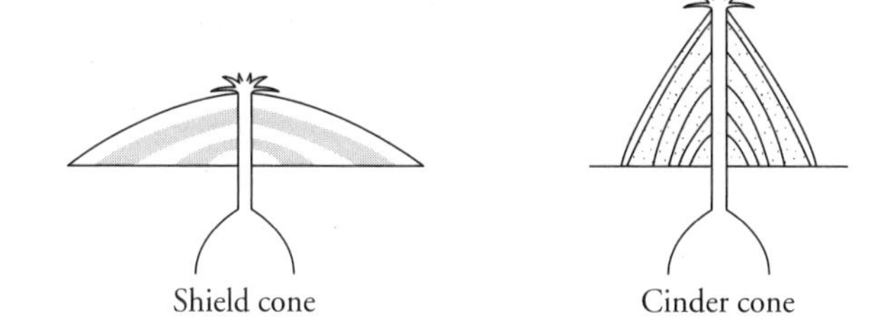

Volcanic Debris

Solid pieces of rock that explode out of a volcano are called tephra. Ash is really pieces of tephra that are less than 2 mm in diameter. Pieces up to 64 mm are called lapilli. Blocks and bombs are fragments that are larger than 64 mm. Tephra can combine with superheated gases to form what is called a pyroclastic cloud (or flow) that moves outward and downhill from the volcano. Sometimes the cloud can move at 100 km/hr. Toxic gases, which can suffocate people, can also be emitted from volcanoes. On the other hand, the erupted volcanic material can produce very fertile soil after it has been weathered.

Types of Eruptions

A **rift eruption** occurs in a long, narrow crack in the Earth's crust. Most of these are located on the ocean floor in areas such as the Mid-Atlantic Ridge and the East Pacific Rise. The lava flows out smoothly in these areas where the plates diverge. At a convergent boundary, where there is an area of subduction, mountains and volcanoes can form and produce **explosive eruptions.** The magma found in these areas is thicker, creating more explosive volcanoes and steeper cones. The Andes Mountains along the western edge of South America and the Cascade Mountains in the western United States contain these types of volcanoes. **Hot spots** are locations where lava flows out but it is the plate that moves. Over a period of time, a chain of mountains is formed. The Hawaiian Islands were formed by a hot spot.

Famous Eruptions

The eruption of Vesuvius in A.D. 79 is an extremely well-known event. The towns of Pompeii and Herculaneum in Italy were buried in ash, which helped to preserve them until they were uncovered. Vesuvius was created by the subduction of the African plate under the Eurasian plate. On August 27, 1883, Krakatoa erupted explosively. Half of the Indonesian island where it was located was blown away. Windows were broken 150 km (90 mi) away from the blast and the sound of the explosion was heard 3,000 km (1,800 mi) away in Australia. Ash reached into sky for 30 km (18 mi) and affected weather patterns in New England. Mount St. Helens in the state of Washington erupted in 1980 and was well documented. Mudflows and ash erupted out with great force. Trees were felled for 25 km (15 mi), ash was up to 5 cm thick, and mudflows combined with landslides to a depth of 180 m as a result of this eruption. Nearby, but 7,000 years earlier, the **caldera** of Crater Lake was formed in Oregon. The volcano, called Mt. Mazama, exploded and has since filled in with rain and snowmelt.

Other Effects from Volcanoes

Besides eruptions of lava flows, gases, and ashes, volcanoes can produce other features. **Hot springs** are formed when groundwater is heated by magma and forced up to the surface, where it escapes through an opening in the ground. If the water deeper below the surface is turned to steam, a **geyser** can result. The steam forces its way to the surface when enough pressure is created. People can harness the energy that heats up the water in hot springs and geysers. This geothermal energy can heat homes and run electric power plants.

Chapter Checkout

Q&A

1. Many volcanoes are found near

a. valleys.
b. glaciers.
c. earthquake zones.
d. areas of erosion.

2. Which of the following usually is not erupted by a volcano?

- **a.** ash
- **b.** gases
- **c.** lava
- **d.** sandstone

3. An example of a hot spot is

- **a.** Vesuvius.
- **b.** Hawaii.
- **c.** Mount St. Helens.
- **d.** Crater Lake.

4. Magma that spreads out horizontally between existing rock layers forms an intrusion called a

- **a.** dike
- **b.** vent
- **c.** batholith
- **d.** sill

Answers: 1. c **2.** d **3.** b **4.** d

Chapter 13
MOUNTAIN BUILDING

Chapter Checkin

- ❑ Determining the type of plate collision
- ❑ Knowing features caused by mountain-building forces
- ❑ Understanding how continents form

The Earth is constantly recycling water, air, and rock materials. As rocks are weathered and eroded or melted in some areas, new rocks are being formed somewhere else. Some of these rocks will turn into mountains.

Types of Crustal Movement

Mountains can be a result of plates colliding. A plate boundary or **active continental margin** is an area where the plates are moving, creating mountains. A **passive continental margin** is a region of quiet water where sediments are being deposited. This will eventually become a mountain as the sediments build. These sedimentary rocks then are lifted upward, forming a mountain.

Plate Collisions and Mountain Types

As plates collide, several scenarios can result, depending on the types of plates involved. When an oceanic plate collides with a continental plate, the more dense oceanic plate of basalt is subducted under the less dense continental plate made of granite. This creates a mountain range parallel to the coast, with some of the mountains being volcanic in nature. A trench is also formed off of the coast.

The collision of two continental plates results in a crumpling of the land. **Folded mountains** are formed. An example of these are the Himalayan Mountains between China and India. These mountains are still growing. The Alps were also formed this way through the collision of Italy (on the

African Plate) and the Eurasian Plate. **Fault-block mountains** are made when the parallel layers of sedimentary rocks are tilted upward. You can see these clearly as you drive through road cuts. If the rock layers are tilted so much that they turn upside down, the rocks are considered to be **overturned.** Magma pushing upward, but not reaching the surface, can create domed mountains. These are generally more rounded than the other types described here. Figure 13-1 shows cross-sections of these mountain types.

Figure 13-1 Mountain cross-sections.

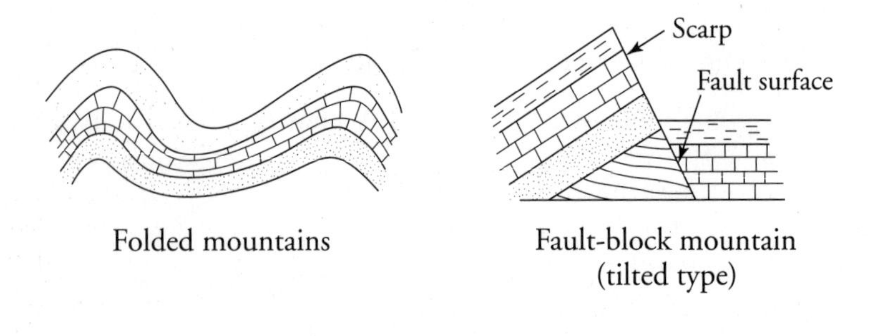

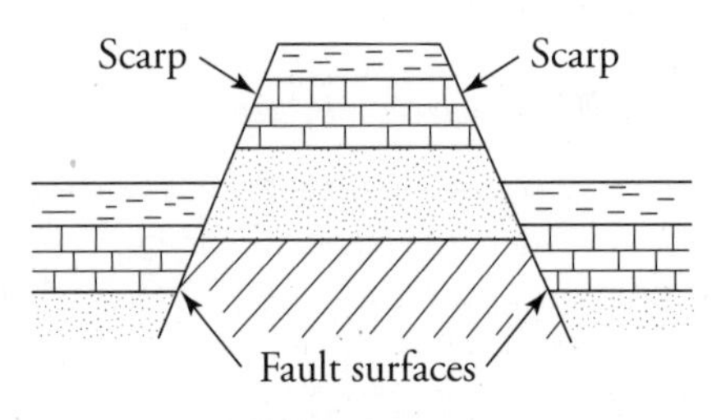

Fault Types and Mountain Features

A break or crack that forms in rocks is called a **fault.** As the rocks are put under tremendous stress, they can break, creating an earthquake. The surface that the rocks move along is known as the **fault plane.** There are three types of faults. A **normal fault** occurs when the rocks on one side move downward. This is usually caused when the rocks are pulled apart. This forms a cliff that could be climbed or hiked upon. If the rocks are pushed toward each other, a **reverse fault,** also called a thrust fault, is formed. The rocks on one side of the fault are pushed upward in relation to the rocks on the other side and an overhanging cliff is formed. The last type is the **strike-slip fault**. This occurs when the two rocks move horizontally to each other. This side-to-side motion can be found along the San Andreas Fault in California. Figure 13-2 compares the different types of faults.

Figure 13-2 Fault types.

If the rocks don't break, but bend instead as is found in folded mountains, other features can be made. Layers of rock folded upward (think of a rainbow) form an **anticline**. A **syncline** forms when the rock layers are folded downward. The rock layers between anticlines and synclines are called limbs. A way to remember the difference between these formations is that the anticline is shaped like the letter *A* and a syncline is shaped like the top of the letter *Y*. Figure 13-3 shows these forms.

Figure 13-3 Anticline and syncline formations.

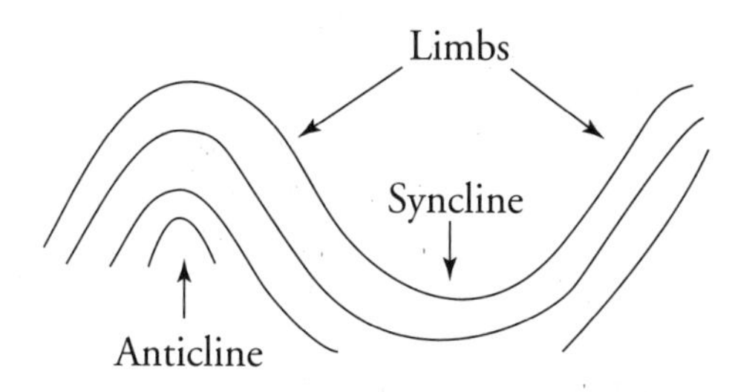

Other Features of the Mountain-Building Process

In addition to making peaks that we can see, there are other features that are evidence of the mountain building process. As uplifting occurs, some whole areas can be lifted together. Raised beaches are formed as a result of this event. Fossils of sea creatures found in the tops of mountains are an indication that the sedimentary rock, which contains these fossils, was formed in the ocean.

Correlation of Mountains and Plate Boundaries

The connection between plate boundaries and the formation of mountains is direct. When plates collide, mountains are created. Examples of these are found all around the world. Some ranges, such as the Himalayas and the Andes, are still growing, whereas others are older and are being eroded faster than they are being created. A few mountainous areas, such as the Appalachians, no longer have the forces in place that created them. These mountains are being weathered and eroded back into the ocean as the Earth recycles its materials.

Growth of Continents

Continents can grow from a variety of events, but have a core rock upon which other material is added. **Cratons** are these cores. The North American craton is located mostly underground. The exposed portion is in eastern Canada around the Hudson Bay area and is called the Canadian Shield. Over time, rivers have deposited material, volcanoes have erupted, and plates have collided to form what we now know as North America. The Appalachian Mountains were formed when the land masses on the Earth were converging. Thin, horizontal sheets of rock from the edge of the continent were pushed inland. The **thin-skinned thrusting** of the layers stacked up to form the Southern Appalachians. The western side of North America was formed when a large piece of rock from another plate was moved over a large distance. This is called a **terrane** and can be identified by several characteristics. The terrane block is surrounded on each side by major faults. The rocks and fossils found in terranes don't match those of nearby rocks. Finally, the magnetic polarity found in the transported block of rock doesn't match the nearby rocks.

Chapter Checkout

Q&A

1. Which is the best evidence of crustal movement?

- **a.** Molten rock in the Earth's outer core
- **b.** Tilted sedimentary rock layers
- **c.** Residual sediments on top of bedrock
- **d.** Marine fossils found below sea level

2. Which statement best explains why sharks' teeth have been found in the bedrock of some mountainous regions?

a. The area was once below sea level.
b. Sharks were once amphibious animals.
c. A type of shark existed on land in early times.
d. Shark remains were transported great distances before being deposited.

3. Which observed feature would provide the best evidence of crustal stability?

a. Horizontal sedimentary layers
b. Changed benchmark elevations
c. Folded, faulted, and tilted strata
d. Marine fossils at elevations high above sea level

4. Which of the following processes don't add material to help continents grow?

a. volcanoes
b. rivers
c. chemical weathering
d. terranes

Answers: 1. b **2.** a **3.** a **4.** c

Chapter 14

ENERGY AND THE EARTH

Chapter Checkin

- ❑ Knowing the ways energy can be transferred
- ❑ Determining the difference between temperature and heat
- ❑ Understanding the phase change diagram for water

The Sun is the ultimate source for all of the energy on Earth. Energy can be transferred by various means and is needed to change the temperature and state of materials.

Transfer of Energy

The transfer of energy drives weather on Earth. Energy can be transferred in four different ways: **radiation, convection, conduction,** and **advection.** The first three are common to other physical-science courses, but advection is unique to meteorology.

Radiation

The Sun transfers its energy through space, which lacks air. The energy is transferred by radiation. This is the same method of energy transfer that allows you to listen to radio stations and see broadcast television programs.

Convection

Energy is transferred in fluids by convection. This includes currents in the ocean and in other liquids and gases. Some heaters work by convection currents. A convection current is created, which starts a circulation in the container that it is in. This continues until the fluid is a uniform temperature.

Conduction

The main method of energy transfer for solid materials is conduction. Energy is transferred when one object or molecule touches another.

Advection

Air is advected when it is moved from one region on Earth to another area. Warm air can be pulled up from the southern United States to a northern location. Usually the Sun raises the air temperature, but advected air can also raise the temperature. Cold air can be pulled down into normally warmer areas. The surface winds and weather systems are responsible for this temperature change.

Absorption of Energy

Three factors affect how an object can absorb energy—color, surface texture, and reflectivity. Black objects absorb energy better than white-colored objects. Smooth objects reflect more energy and radiation than rough objects. The rough surfaces have more surface area to allow for more absorption at different angles. Shiny objects reflect more energy than dull objects. A good absorber of energy is also a good radiator. Poor absorbers of energy make for good insulators.

Types of Energy

The types of energy and resources available on the Earth fall into two categories—renewable and nonrenewable. **Renewable energy** and resources can be replenished in a relatively short amount of time. The recycling of resources helps to lessen the need to take more from the Earth. Renewable energy sources include solar, water power, wind power, and geothermal. Nonrenewable resources are most metals and ores. **Nonrenewable energy** sources are oil, coal, natural gas, and nuclear energy. One of the controversial issues regarding the use of nuclear energy, however, is the disposal of the waste materials. Gasoline made from grains and soy are being investigated as alternative energy sources. Conservation and investigation of alternative energy sources are the keys to ensuring that we will have energy for many generations.

Temperature and Heat

Temperature is the measurement of the average kinetic energy of air molecules. A thermometer is used to measure this energy. Three different scales are used to measure temperature.

The Fahrenheit scale is used mainly in the Untied States. Ironically, it is part of the English measurement system, but England (and most of the world) uses the metric system. Temperatures recorded on station models are still labeled in degrees Fahrenheit. G. Daniel Fahrenheit, a German physicist, developed the Fahrenheit scale in 1724. He set the lower limit of his scale to the lowest temperature he could find outside and the upper end as his body temperature. He was trying to avoid negative numbers with his scale.

The scale for recording temperatures in the metric system is the Centigrade or Celsius scale. It was named for its creator Anders Celsius, a Swedish astronomer, who introduced it in 1742. His scale is based on the freezing point (0°) and boiling point (100°) of water.

A third scale is also used, which combines the two preceding scales. William Thomson, 1st Baron Kelvin, a Scottish-Irish mathematical physicist and engineer, developed this scale in 1848. The scale for degrees is the same as the Celsius scale, except there aren't any negative numbers. 0 K is the temperature **absolute zero.** At this point, all motion of molecules and atomic particles ceases. Above this temperature, all matter gives off heat energy. The amount of heat taken in and given off depends on the temperature and composition of the substance. The three different scales are compared in Table 14-1.

Table 14-1 Temperature Scales

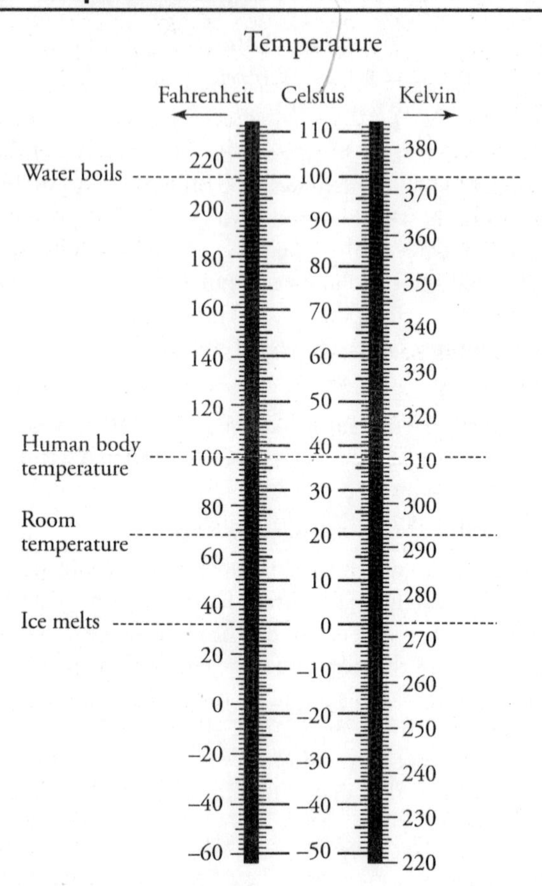

Change in State of Water and Matter

Water is the only material that exists as a solid, liquid, and gas on the Earth at any given time. There is snow and ice at the poles as well as elsewhere. Liquid water covers the majority of the Earth's surface. Water vapor is present in the air we breathe, even in desert areas.

Conversions between States

In order for water to change state, energy is needed. This energy is added or taken away. The processes to go from solid to liquid (melting) and liquid to gas (boiling or **vaporization**) require energy to be added. The reverse procedures, going from gas to liquid **(condensation)** and liquid to solid (freezing), give off energy. This released energy is important in helping tropical storms and hurricanes develop and intensify. **Deposition** is the process that can occur when water goes from the gas state directly to the solid state. **Sublimation** occurs when a solid turns into a gas. Frozen carbon dioxide (dry ice) is an example of this. The outcome of deposition is snow. The visual difference between snow and ice can help determine how each was formed. Ice is clear, whereas snow is white. The process that causes snow to form rearranges the water molecules so that light can reflect off of them, as opposed to ice, which is transparent.

Energy Required for Change of State

The amount of energy required to change water between the solid and liquid states is much less than the energy needed to go between the liquid and gas states. Table 14-2 shows the energy needed for these energy changes.

Table 14-2 Properties of Water

Energy gained during melting	80 calories/gram
Energy released during freezing	80 calories/gram
Energy gained during vaporization	540 calories/gram
Energy released during condensation	540 calories/gram

During the time of a phase change, the temperature of the water doesn't change. The energy being added helps in the conversion from one state to another. After all of the water molecules have changed state, the temperature then begins to rise again. Figure 14-1 shows the phase change curve for water.

Figure 14-1 The phase change curve for water.

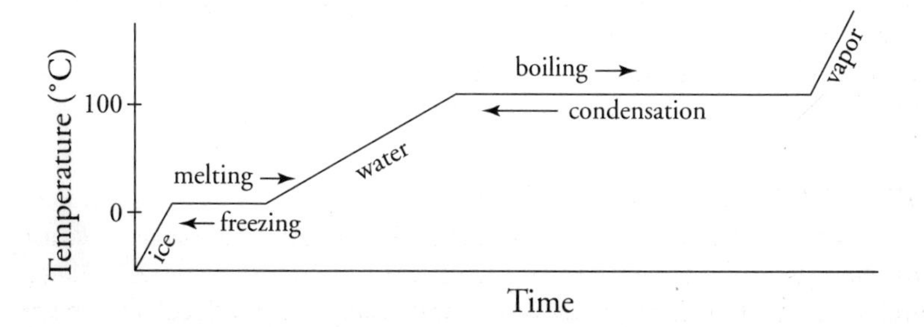

Energy of the Earth

The energy that radiates out from the Earth comes from several sources. Heat was generated when the Earth was initially formed. The decay of radioactive elements in the Earth provides a lot of heat. The last main source is friction from the rubbing of plates caused by plate tectonics in the Earth's crust.

Chapter Checkout

Q&A

1. Which is the major source of energy for most Earth processes?
   - **a.** Radioactive decay within the Earth's interior
   - **b.** Convection currents in the Earth's mantle
   - **c.** Radiation received from the Sun
   - **d.** Earthquakes along fault zones
2. What is the primary method of heat transfer through solid rock during contact metamorphism?
   - **a.** advection
   - **b.** convection
   - **c.** absorption
   - **d.** conduction
3. The Earth loses heat energy to outer space mainly by
   - **a.** radiation
   - **b.** reflection
   - **c.** convection
   - **d.** conduction
4. Which type of surface would most likely be the best reflector of electromagnetic energy?
   - **a.** light-colored and smooth
   - **b.** light-colored and rough
   - **c.** dark-colored and smooth
   - **d.** dark-colored and rough

Answers: 1. c **2.** d **3.** a **4.** a

Chapter 15

INSOLATION AND SEASONS ON EARTH

Chapter Checkin

- ❑ Understanding when the warmest times of the day and year occur
- ❑ Determining the position of the Earth at the beginning of each season
- ❑ Knowing how the greenhouse effect works

The seasons on the Earth are not caused by how close the Earth is to the Sun. The Earth is closest to the Sun on or about January 1, and farthest away on or about July 1 each year. The reason for the seasons lies in the amount of the Sun's radiation that reaches the Earth.

Solar Radiation

The amount of energy put out by the Sun is a constant. The incoming solar radiation is known as **insolation.** The amount of solar energy reaching the Earth is 70 percent. The surface of the Earth absorbs 51 percent of the insolation. Water vapor and dust account for 16 percent of the energy absorbed. The other 3 percent is absorbed by clouds. Of the 30 percent that is reflected back into space, 6 percent is reflected by air and dust. Clouds reflect 20 percent, and the remaining 4 percent is reflected by the surface. The energy that was absorbed can be reradiated. Of the reradiated energy, 70 percent is lost to outer space. The surface is responsible for 21 percent of this, and the remaining 49 percent is lost by the atmosphere. The remaining 30 percent is transferred by the surface to the atmosphere.

Changes in Insolation

The amount of insolation that an area receives can vary over the course of a day or over a year. The highest point of the Sun's path in the sky is the time when the maximum amount (intensity) of insolation for the day reaches a location. The warmest part of the day is usually a few hours later. This is because the land absorbs the sunlight and reradiates it out to the atmosphere, warming it up. We measure the air temperature when giving temperature readings. The coldest part of the day is just before sunrise, when the Earth's surface has reradiated much of the energy it absorbed during the hours of sunlight.

Over the course of a year, the Sun reaches its highest point on June 21 for anyone living north of the Tropic of Cancer. The maximum air temperature for this area is delayed until July. The reason is similar to the daily changes. The ground needs time to absorb the energy and to reradiate it to the atmosphere. On the other end of the year, the Sun reaches its lowest noontime point on December 21. For the same reasons as above, the coolest month for the region is January.

Greenhouse Effect

Visible energy that comes in from the Sun is a short wavelength. This is absorbed and reradiated as a longer wavelength, such as infrared. This is because energy is lost during the transition. These longer waves are heat waves. If the reradiated energy is trapped, a greenhouse situation is created. The trapping mechanism can be clouds or greenhouse gases, as is the case on Venus. Carbon dioxide gas in the atmosphere acts like the glass in a hothouse. Basically, the Sun's rays check in, but they don't check out. A greenhouse situation is shown in Figure 15-1. Earth's main greenhouse gases are carbon dioxide, water vapor, and methane.

Figure 15-1 Greenhouse effect.

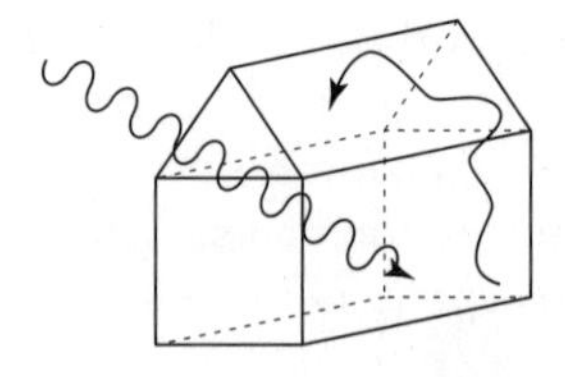

Seasons

The seasons on Earth are caused by the 23.5° tilt of the axis of rotation. The summer begins on or about June 21, when the Sun is directly overhead at local noon on the Tropic of Cancer (23.5°N latitude). This is the **Summer Solstice.** Solstice means "Sun stands still" in Latin. All points above the Arctic Circle (66.5°N latitude) have 24 hours of sunlight; above the Antarctic Circle (66.5°S) all points have 24 hours of darkness. In the United States, the Sun rises north of due east and sets north of due west. As it reaches the noon position, it is in the southern part of the sky. For an observer in the United States, the Sun is never directly overhead. The opposite occurs at about December 21, the first day of winter **(Winter Solstice).** The Sun is directly overhead at the Tropic of Capricorn (23.5° S latitude). The North Pole experiences total darkness, whereas the South Pole is in total light. For an observer in the United States, the Sun rises south of due east and sets south of due west.

The equinoxes ("equal night") fall on or about March 21 **(Vernal Equinox)** and September 23 **(Autumnal Equinox).** The Sun is directly overhead at the Equator and the entire Earth has 12 hours of day and night. The Sun rises directly in the east and sets directly in the west. Contrary to a popular myth, you can balance an egg on its end on any day of the year, not just when the day and night are in balance. It is just a function of gravity. Figure 15-2 shows the positions for each season.

Figure 15-2 The positions of the Earth at the start of each season.

Chapter Checkout

Q&A

1. For which location and date will the shortest duration of insolation occur?
   - **a.** 60°N latitude on June 21
   - **b.** 23.5°N latitude on June 21
   - **c.** 60°N latitude on December 21
   - **d.** 23.5°N latitude on December 21
2. In New York State, the longest period of insolation occurs on or about
   - **a.** March 21.
   - **b.** January 21.
   - **c.** August 21.
   - **d.** June 21.
3. A student in New York State observed that the noon Sun increased in altitude each day during the first part of a certain month and then decreased in altitude each day later in the month. During which month were these observations made?
   - **a.** November
   - **b.** September
   - **c.** June
   - **d.** February
4. Scientists have theorized that an increased concentration of carbon dioxide will cause an increase in worldwide atmospheric temperature. This theory is based in the fact that carbon dioxide is a
   - **a.** good absorber of infrared radiation.
   - **b.** poor absorber of infrared radiation.
   - **c.** good reflector of ultraviolet radiation.
   - **d.** poor reflector of ultraviolet radiation.

Answers: 1. c **2.** d **3.** c **4.** a

Chapter 16
WEATHER

Chapter Checkin

- ❑ Knowing the structure and composition of the atmosphere
- ❑ Determining the dew point, relative humidity, and cloud base for air
- ❑ Understanding how mid-latitude cyclones form

Weather is the state of the atmosphere at a given time and place. Different measurements can be made to determine the condition of the atmosphere. The primary variables are barometric pressure, temperature, humidity, precipitation, wind speed, and wind direction. Meteorologists study these readings to help forecast the weather.

Composition of the Atmosphere

Most of the atmosphere is composed of nitrogen. In addition to 78 percent nitrogen, 21 percent is oxygen. The remaining 1 percent is composed of argon, carbon dioxide, and other gases. 99 percent of the atmosphere is found within 32 km of the surface of the Earth. Half of the atmosphere's weight is found from the surface up to 5.5 km in the atmosphere. Other particles found in the atmosphere include water vapor, ozone, and dust. Ozone (O_3) is found at 10 to 50 km above the Earth's surface and is very important to life on Earth. It absorbs harmful UV radiation, but is being destroyed by CFCs (chlorofluorocarbons) in human products, such as aerosol sprays. The dust found in the atmosphere is made from rock and mineral particles, pollen, sea salt, chemicals, and bacteria. These dust particles cause hazy skies and provide a surface for condensation of water vapor.

Temperature in the Atmosphere

As you rise up through the atmosphere, the temperature can vary greatly. The lowest level is the **troposphere,** which starts from the surface of the Earth. This level reaches to about 8 km high at the poles and 18 km high at the equator. The temperature gradually decreases to about –55°C until the **tropopause** is reached. This is the level where weather occurs. After passing through the tropopause, which is just a transition zone, the **stratosphere** starts. Here the temperatures rise to about 0°C at the **stratopause.** This is about 50 km above the Earth. Above the stratopause is the **mesosphere.** The temperatures begin to fall again to about –90°C at the **mesopause,** which is about 80 km in altitude. The **thermosphere** is the uppermost layer of the atmosphere, and the temperatures again begin to rise here. Because the air is dry, the temperatures can rise to over 100°C. The thermosphere continues upward to about 500 km above the Earth's surface.

The Ionosphere

The ionosphere is a region of the atmosphere ranging from 65 km above the Earth's surface to its outer edge at an altitude of 500 km. This region has an abundance of ions that were formed from incoming solar ultraviolet radiation. The ions reflect radio waves back to Earth. During the day, the ion layer is lower in the atmosphere due to the Sun's radiation. At night, the layer moves up higher in the atmosphere. This helps radio waves, especially AM radio, to be reflected over longer distances across the Earth. The ionosphere is affected by solar flares and the 11-year sunspot cycle. The years of maximum activity in the sunspot cycle suppress the ionosphere. Auroras are also formed when these charged particles reach the Earth and interact with the magnetic field.

Moisture in the Atmosphere

The troposphere is also known as the weather sphere. This is due to the water vapor in the air. After the tropopause, water vapor doesn't exist in the atmosphere. Figure 16-1 shows the temperature, pressure, and water vapor values for each of the layers of the atmosphere.

Figure 16-1 Cross section of Earth's atmosphere.

Selected Properties of Earth's Atmosphere

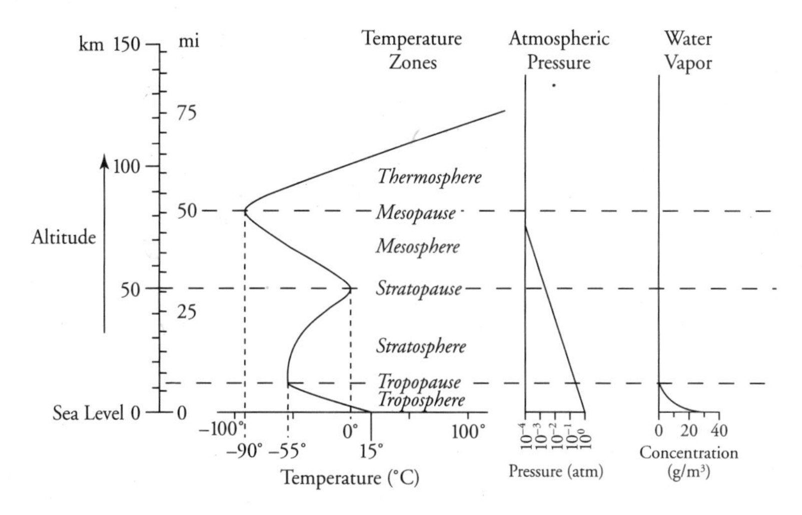

Measurement of Water in the Air

The amount of water in the air can be measured in different ways. The **specific humidity** of air is a measure of how much water is in the air. Warmer air can hold more water than colder air. When the air reaches its capacity, it is **saturated.** This capacity doubles for about every 11°C rise in temperature. The term more often used is **relative humidity.** This is the measure of how much water is in the air divided by how much it can hold. The relative humidity reading is given as a percent. The relative humidity for saturated air is 100 percent.

Finding Relative Humidity

The relative humidity can be found by two different methods. One involves the use of a **hygrometer.** This is a pointer attached to a piece of hair. As the humidity increases, the hair stretches out. This is your typical "bad-hair day." When the humidity drops, the hair shrinks, causing the needle to point in a different direction. The other method requires the use of two thermometers and a chart. The thermometers and chart all use the Celsius scale. If you are getting the readings from a station model and need to find the relative humidity, you need to convert the temperature readings from Fahrenheit to Celsius. Station models show the surface observations and weather data for a specific city. One of the thermometers measures the air temperature. This is the **dry-bulb** reading. The other thermometer has a wet wick on the bottom of the bulb. Water evaporating from the wick into the air takes energy with it, cooling off the thermometer. As the relative humidity increases, less water can evaporate into the air. This makes the temperature readings between the two thermometers closer. If the air is much drier, the thermometers have readings that are much farther apart. The difference between the wet and dry-bulb temperatures is called the **wet-bulb depression.** The wet-bulb temperature is always lower or the same as the dry-bulb temperature. If the temperatures are the same, the relative humidity is 100 percent. The wet-bulb depression is used with the dry-bulb temperature and a chart to determine the relative humidity. Table 16-1 shows this chart.

Table 16-1 Chart for Determining Relative Humidity (%)

Relative Humidity (%)

Dry-Bulb Temperature (°C)	Difference Between Wet-Bulb and Dry-Bulb Temperatures (°C)															
	0	1	2	3	4	5	6	7	8	9	10	11	12	13	14	15
–20	100	28														
–18	100	40														
–16	100	48														
–14	100	55	11													
–12	100	61	23													
–10	100	66	33													
–8	100	71	41	13												
–6	100	73	48	20												
–4	100	77	54	32	11											
–2	100	79	58	37	20	1										
0	100	81	63	45	28	11										
2	100	83	67	51	36	20	6									
4	100	85	70	56	42	27	14									
6	100	86	72	59	46	35	22	10								
8	100	87	74	62	51	39	28	17	6							
10	100	88	76	65	54	43	33	24	13	4						
12	100	88	78	67	57	48	38	28	19	10	2					
14	100	89	79	69	60	50	41	33	25	16	8	1				
16	100	90	80	71	62	54	45	37	29	21	14	7	1			
18	100	91	81	72	64	56	48	40	33	26	19	12	6			
20	100	91	82	74	66	58	51	44	36	30	23	17	11	5		
22	100	92	83	75	68	60	53	46	40	33	27	21	15	10	4	
24	100	92	84	76	69	62	55	49	42	36	30	25	20	14	9	4
26	100	92	85	77	70	64	57	51	45	39	34	28	23	18	13	9
28	100	93	86	78	71	65	59	53	47	42	36	31	26	21	17	12
30	100	93	86	79	72	66	61	55	49	44	39	34	29	25	20	16

The instrument with the wet and dry-bulb thermometers is called a **psychrometer.** A **sling psychrometer** is a handheld device with the two thermometers that spins around.

Dew Point

The temperature at which water vapor condenses into liquid water is called the **dew point.** If the dew point is below 0°C, it is called the **frost point.** These are important numbers in helping to determine where to post frost and freeze warnings. The dew-point temperature is found in a similar manner to that of relative humidity. The dry-bulb and wet-bulb temperatures are determined. The wet-bulb depression and dry-bulb temperature are used with the chart in Table 16-2 to find the dew point.

Table 16-2 Chart for Determining Dew-point Temperature (°C)

Dew-point Temperatures (°C)

Dry-Bulb Temperature (°C)	Difference Between Wet-Bulb and Dry-Bulb Temperatures (°C)															
	0	1	2	3	4	5	6	7	8	9	10	11	12	13	14	15
–20	–20	–33														
–18	–18	–28														
–16	–16	–24														
–14	–14	–21	–36													
–12	–12	–18	–28													
–10	–10	–14	–22													
–8	–8	–12	–18	–29												
–6	–6	–10	–14	–22												
–4	–4	–7	–12	–17	–29											
–2	–2	–5	–8	–13	–20											
0	0	–3	–6	–9	–15	–24										
2	2	–1	–3	–6	–11	–17										
4	4	1	–1	–4	–7	–11	–19									
6	6	4	1	–1	–4	–7	–13	–21								
8	8	6	3	1	–2	–5	–9	–14								
10	10	8	6	4	1	–2	–5	–9	–14	–28						
12	12	10	8	6	4	1	–2	–5	–9	–16						
14	14	12	11	9	6	4	1	–2	–5	–10	–17					
16	16	14	13	11	9	7	4	1	–1	–6	–10	–17				
18	18	16	15	13	11	9	7	4	2	–2	–5	–10	–19			
20	20	19	17	15	14	12	10	7	4	2	–2	–5	–10	–19		
22	22	21	19	17	16	14	12	10	8	5	3	–1	–5	–10	–19	
24	24	23	21	20	18	16	14	12	10	8	6	2	–1	–5	–10	–18
26	26	25	23	22	20	18	17	15	13	11	9	6	3	0	–4	–9
28	28	27	25	24	22	21	19	17	16	14	11	9	7	4	1	–3
30	30	29	27	26	24	23	21	19	18	16	14	12	10	8	5	1

If the water vapor comes in direct contact with the cooler surface, it can condense onto it. Fog can occur when warm air moves into an area that has a cold surface temperature. An **advected fog** forms in this situation.

This also can be a reverse situation, where cooler air moves over a warmer surface. **Ground fog** forms by radiational cooling at night. These are common in humid valleys and near rivers and lakes.

Clouds

A cloud is formed when air is cooled to its dew-point temperature. The air cools as it rises away from the Earth's surface. If that temperature is above 0°C, the cloud is made of water droplets. If the cloud forms below 0°C, the cloud is made from ice and snow crystals and supercooled water.

Cloud formations fall into three categories. **Cirrus** clouds are very high clouds that are made from ice crystals. They are the thin, feathery clouds you see on a nice day. **Stratus** clouds are the layered, sheet-like clouds. They are found at lower altitudes. **Cumulus** clouds are the puffy, cottonlike clouds formed by vertical rising of air. Other clouds are made from combinations and variations of these clouds. The name of a cloud may also contain a prefix or suffix that tells you more about the cloud. Alto (high) and nimbus (rain) are some examples of these.

Cloud Development

As a parcel of air rises upward, it cools. The air expands and cools because of the decreasing pressure. The rate at which it cools depends on the amount of moisture in the air. If dry air rises, it cools at a rate of 1°C/100 m. This is the **dry adiabatic lapse rate.** By adding moisture, this rate changes to 0.6°C/100 m. This is the **moist adiabatic lapse rate.** The high specific heat of the water is the reason for the difference in the rates. When air at the surface is heated, it rises upward. The air is warmer than the air surrounding it and is less dense, which makes it **buoyant.** This is why clouds appear to "float" in the sky. The clouds can continue to develop vertically. Eventually, a cumulonimbus cloud may form. These are thunderstorm clouds that can be associated with heavy rain, hail, strong winds, and tornadoes. These clouds form in an unstable air mass that has air that is moving due to density differences.

A cloud can form in a stable air mass, but it rises for other reasons. These are layered clouds that form from air that is forced upward by the land (mountains) or by radiational cooling as the air mixes with a cooler layer of air. Some clouds that form have a flat base and billow out on top. The bottom of the cloud is the place where the air temperature is the same as the dew-point temperature. This is the point known as the condensation level. The height of the cloud base can be found with a simple formula or a chart. To use the formula, take the difference between the temperature and dew point at the surface and divide it by 0.8°C (the amount that the dew-point temperature gets closer to the air temperature in 100 m). The result is multiplied by 100, which gives you the **lifting condensation level** or the height that a cloud can form at. Cloud base altitude can also be found by using the air temperature and the dew point temperature. The air temperature is plotted along the solid lines and the dew point follows along the dashed lines in Figure 16-2. When the lines meet, read along the side that is labeled "Altitude." This is the height of the cloud base in kilometers.

Figure 16-2 Graph for determining cloud base altitude.

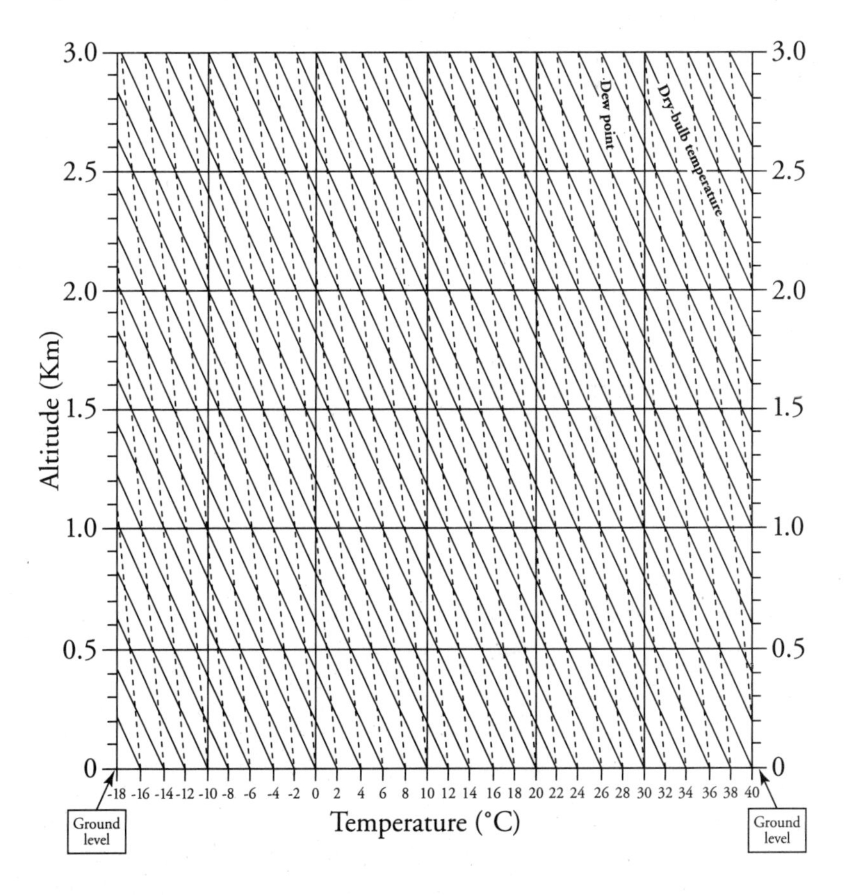

Eventually the cloud and air temperatures become equal. The cloud isn't buoyant at this point and begins to spread out. This creates the classic anvil-shaped tops that are seen at cloud tops.

Precipitation

In order for water vapor to condense, certain conditions are needed. The air must cool down. This can occur in several different ways. It can come in contact with a colder surface; it can radiate heat; it can mix with colder air; or it can expand as it rises upward. The other ingredient that is needed is condensation nuclei. This provides a surface for condensation to occur. These particles can be dust, salt, sulfate, or nitrate particles (these form acid rain) in the air. Scientists have seeded clouds to enhance nucleation and produce needed rain. Silver iodide crystals are put into clouds to provide a surface for condensation to occur. In some instances, water vapor can condense and form water droplets (homogenous nucleation), but this is rare. The type of precipitation that forms depends on the air temperature. If this is above the freezing point, rain forms. If the air temperature is below 0°C, snow forms. Figure 16-3 shows the air conditions needed for different types of precipitation.

Figure 16-3 Air and surface temperatures and resulting precipitation.

Clouds / Surface	+0°C / –0°C	+0°C / +0°C	–0°C / –0°C	–0°C / +0°C
Precipitation	Sleet/ Freezing Rain	Rain	Snow	Rain

Updrafts in a cloud move rain droplets around. As they collide, they collect and get larger. When the drop gets too heavy to stay in up the cloud, it falls to the Earth. Small droplets of rain (less than .02 cm in diameter) are called **drizzle.** Drops larger than these are called **rain.** When much larger drops of rain fall, they fall apart into smaller drops by vibrations caused by friction with the air.

Hail forms in a tall cloud with strong updrafts. An ice crystal or frozen raindrop moves though the cloud collecting water droplets. As the hailstone rises up in the cloud, the outer layer freezes. When it falls downward, it gathers more water droplets. This circulation process continues until the hailstone falls to the ground. When you cut a hailstone in half, you can see rings. Like rings in a tree, they can tell the hailstone's history of formation. Some hailstones can reach the size of a softball. These can be very damaging to crops, animals, cars, and other property.

Pressure in the Atmosphere

Air pressure is the weight of the air over a specific area. At sea level this is about 1 kg/cm^2. The pressure gradually decreases from the surface of the Earth at a rate of about 1 cm Hg/123 m (1 in/1,000 ft) in the first few kilometers. About half of the weight of air is found in the first 5.5 km of the atmosphere. The air gets very thin above that point. The layer of air continues until about the middle of the mesosphere. At this point, air is too thin to be measured.

Air pressure is measured using a barometer. The two types are mercury and aneroid. The mercury thermometer is a column of mercury in a tall tube. This tube is inverted into a bowl of mercury. The air puts a force on the mercury in the bowl, keeping the mercury in the tube from flowing out. The height of the mercury column is about 76 cm (30 in) tall. As the pressure increases, it forces mercury higher up into the tube. When the air pressure falls, the level in the tube falls as well. When you hear the barometric pressure reading on a weather report, this is the value that they are referring to. The units used are centimeters (or inches) of mercury. Millibars are used in the metric system and on station models. A millibar is about 1/1,000 of the pressure at sea level. Standard (average) sea-level pressure is 1,013.2 mb (29.92 in of Hg). This is also called 1 atmosphere of pressure. Figure 16-4 shows the pressure scales.

Figure 16-4 Scales showing pressure in millibars and inches of mercury.

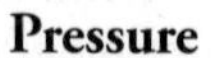

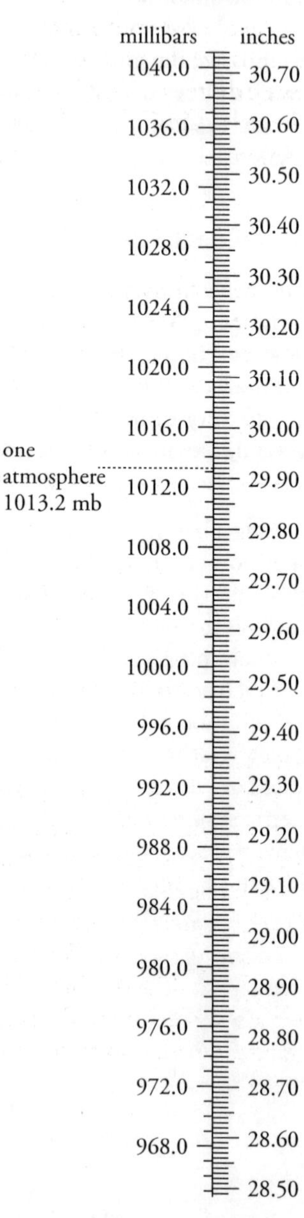

An aneroid barometer is a sealed container with a pointer attached to it. As the pressure increases, it squeezes the container. This moves the pointer in one direction. The pointer moves in the reverse direction as the pressure falls. Air pressure can change for several reasons. The amount of moisture content in the air can change the pressure. As the amount of water in the air increases, the pressure decreases. This is because a molecule of water has less mass than a molecule of air. Cold air is denser than warm air and therefore has a higher pressure.

Wind

The winds at the surface of the Earth vary depending on the location. In the stratosphere, the winds are steady. The air is clear and dry there. The consistency of the airflow allows airplanes to fly on regular courses. Wind speed is measured using an **anemometer.** The direction is shown by a wind vane that points into the direction that the wind is coming from. This is how winds are named. On the station model, the wind speed is recorded in knots (nautical mi/hr). A knot is about 1.15 mi/hr (1.85 km/hr).

Causes

The differences in pressure cause air to move. The larger that this gradient is, the stronger the air flow. The air wants the pressure to be equal, but different factors cause the air pressure to differ. The main cause is unequal heating of the Earth. As air is heated, the pressure decreases.

Circulation of Air on Earth

The unequal heating on the Earth causes air to rise and sink in different regions. If the Earth didn't rotate, the air would rise along the equatorial regions and sink at the poles. Because the Earth rotates, the circulation of air around the Earth is more complex. The Coriolis Force causes the winds to bend to the right in the Northern Hemisphere and to the left in the Southern Hemisphere. The strength of the Coriolis Force depends on latitude. As the latitude increases, so does the force. The amount of land and water in each hemisphere also causes more differences in heating. All of these factors cause the global circulation to be made up of several cells. These are seen in Figure 16-5.

Figure 16-5 Planetary wind and moisture belts in the troposphere.

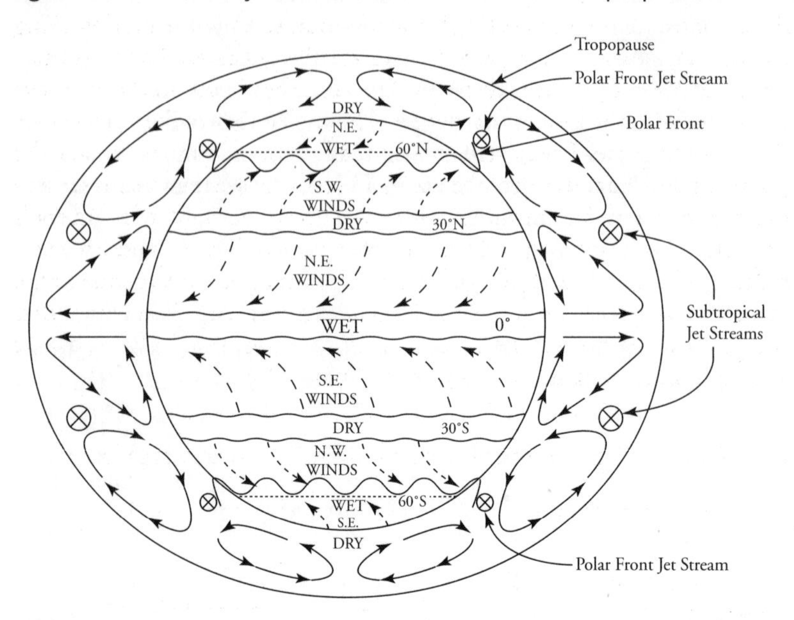

Along the 0° and 60° latitude lines are areas where the air is rising. These are areas of low pressure, which leads to more rainy conditions. The 30° and 90° latitudes are regions where the air is sinking. This creates areas of higher pressure and drier air. These bands move north and south as the concentration of the Sun's rays migrates over the course of a year. The surface winds along each of the bands just mentioned are light, whereas the areas between these zones have constant winds.

Some of these areas have been named. The area along the equator has the same weather conditions for most of the year and is called the **doldrums.** Along the 30° latitude line, the winds die off. This caused problems for sailors in the Northern Hemisphere who were bringing horses to the New World across the Atlantic Ocean in the 1700s. These horses were considered dead weight and thrown overboard to lighten the ships, allowing them to sail through to their destinations. Thus, these are known as the **Horse Latitudes.** Between 0° and 30° are the **trade winds.** These winds constantly blow from the northeast. Most of the United States falls in the area between the 30° N and 60° N band. The winds here blow from the southwest and are called the **prevailing westerlies.** These winds help to guide weather systems.

Sea Breeze/Land Breeze

A local effect of uneven heating is the **sea breeze.** Daytime heating along a beach area warms the land and water at different rates. The land heats up much faster than the water does. The land then heats up the air above it. The air becomes less dense and rises. The cooler air over the water moves in to take its place. The heated air eventually cools and moves to take the place of the air that was over the water. This convection cell that is created causes a sea breeze for anyone who is on the beach. At night, the land cools off faster than the water and the air reverses direction. A **land breeze** is formed. The land breezes are generally weaker than sea breezes due to the fact that water cools off more slowly than land heats up. Remember that the winds are named for the direction that they came from. The sea breeze and land breeze situations are shown in Figure 16-6.

Figure 16-6 Circulation patterns for sea breezes and land breezes.

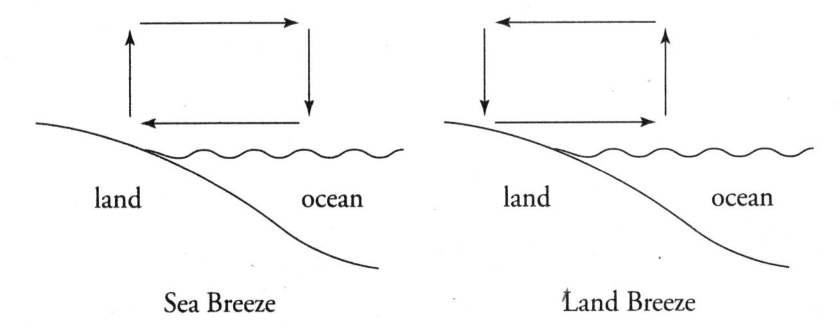

Wind Shifts

For the same reason that sea and land breezes form, larger-scale winds are created between continents and oceans. In the summertime, the oceans are cooler than the continents. The areas of cooler, sinking air over the oceans cause high pressure to be formed. Areas of lower pressure are formed over the land. Along the east coast of the United States, a Bermuda High forms, creating the hot, humid conditions that are experienced. The reverse happens in the wintertime, when the land is colder than the ocean. In some areas of the world, seasonal winds are created. These **monsoon** winds cause wet and dry seasons. In India, the winds coming off of the Indian Ocean create extended periods of heavy rain. When the winds shift and come down from the Himalayan Mountains, much drier conditions are experienced. Near the top of the troposphere, about 6 to 12 km (3.5 to 7 mi) up from the surface, is a band of very fast-moving air. This is the **jet stream,** and it moves at about 300 to 500 kmph (180 to 300 mph). This river of air helps to move weather systems in the lower parts of the troposphere. The jet stream also affects the courses of airplanes.

Air Masses

A huge area of air in the lower section of the troposphere that has similar characteristics throughout is called an **air mass.** Air masses are named based on their characteristics. These variables are the temperature and moisture content. Air masses coming from colder areas are labeled as polar (P), whereas tropical masses (T) come from warm regions. Extremely cold regions supply arctic (A) air masses. If the source region is over the ocean, it is moist and labeled as maritime (m). Air originating from over land is called continental (c). The combination of the five labels and conditions describes air masses. For example, an mT air mass could originate in the Gulf of Mexico, whereas a cP air mass could come from the middle of Canada. The other possible air masses are cA (continental arctic), cT (continental tropical), and mP (maritime polar). As the masses move along into other areas, their characteristics can moderate. The weather experienced while in a particular air mass doesn't change much. The clashing of air masses creates other weather conditions that will be discussed in the next section.

Fronts and Mid-latitude Cyclogenesis

When two different air masses meet, a **front** is formed. The boundary between these masses is called a **stationary front.** The air between the two masses never really combines. The stationary front may only last for a few days. One of the air masses, usually the colder and drier mass, pushes to the south, while a warmer and moister air mass moves up from the south. Because of the Coriolis Force, the masses start to move to the right of the direction that they were originally moving. A counterclockwise rotation is started. A **low-pressure center** forms along the stationary front. As the system spins and moves to the east (in the United States), a **cold front** and a **warm front** are created. The cold front is the leading edge of the cold air and is fast moving. The cold air acts as bulldozer, pushing the warmer air. The air in the warmer sector rises and forms clouds.

Thunderstorms and other violent storms are characteristic along cold fronts. They are usually quick-moving storms. When the cold front passes, the winds shift from a southerly direction to the northwest. The air also becomes drier. As the warm air moves along, the front edge of it tries to move the colder, denser air ahead of it. It cannot and rides up the "bubble" of air. As the warm air rises, it cools and forms clouds. Because the warm front is slow moving, it can cause several days of steady rain. The air at the surface is cool, making for dreary days. As the warm front passes, the air temperature rises and the winds come from a southerly direction. Eventually the cold front catches up to the warm front, squeezing the air of the warm sector upward. This forms an **occluded front,** and usually precipitation falls on the Earth below. This is near the end of the life cycle of the low-pressure center. These are called **mid-latitude lows** and the process is called **cyclogenesis.** The side view of fronts, labels for fronts, and steps of cyclogenesis can be seen in Figure 16-7.

Figure 16-7 Side view of fronts.

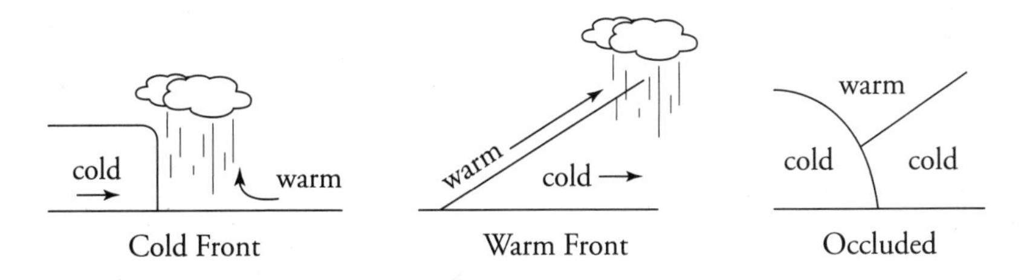

Mid-latitude cyclogenesis

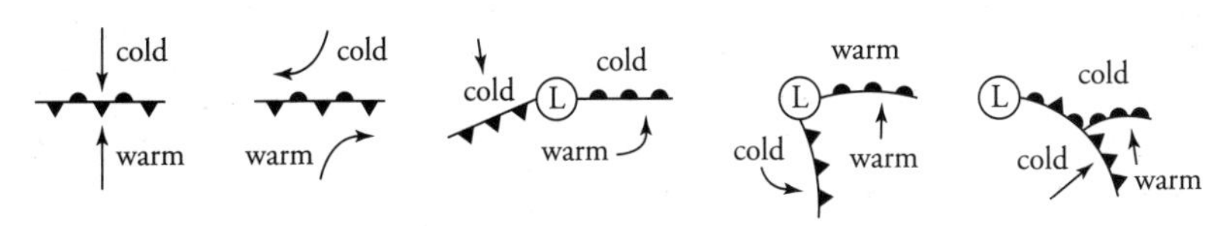

Front symbols*

cold

warm

stationary

occluded

*front symbols point in the direction that the air is moving

The winds around a low-pressure center move inward and counterclockwise. This is the opposite for a high-pressure system, where the winds move outward from the center and clockwise. The winds are generally lighter in highs and the skies are generally clear.

Storms and Severe Weather

Some low-pressure systems can give rise to weather events that are damaging to people, their property, and the land. These incidents can occur almost anywhere and at any time of the year. If the conditions are right for a severe storm to occur, the National Weather Service posts a watch. This means that it is possible to experience the event that the watch is posted for. If the storm is sighted and on its way to you, a warning is posted.

Thunderstorms

Across the globe, about 44,000 thunderstorms occur daily. Although most form in warmer conditions, some can be found during the fall and winter. Many of these thunderstorms form along the **Intertropical Convergence Zone** (ITCZ). This is the region found at about the equator where the trade winds from the Northern Hemisphere and the Southern Hemisphere meet. The air converges and rises upward. This lifting of the air causes these storms to form. This band of the ITCZ moves north and south over the course of a year with the **vertical ray** of the Sun.

A **thunderstorm** can form in other areas and by different means. A local air mass can produce enough lifting of the air to form a thunderstorm. Many thunderstorms are formed along cold fronts. The air along the front is lifted quickly to help in its formation. There may be squall lines caused by strong turbulence ahead of the front. These can create thunderstorms.

Formation

Thunderstorms are formed by air rising upward. Air is lifted by the daytime heating of the Earth's surface. The air rises and cools, forming a cloud. This cloud continues to build taller, causing strong updrafts and downdrafts. The updrafts supply the energy for the storm. Rain and possibly hail are produced and fall to the ground. This downward movement of the precipitation eventually cancels out the updrafts and the storm dies off. If the top of the cloud forms an anvil shape, the rain could follow the clouds outward and fall to the ground without going through the core of the storm. These **supercells** can exist for hours and spawn tornadoes. A severe thunderstorm has winds in excess of 50 mph and can produce hail of 2 cm or more in diameter.

Lightning

Besides strong winds and heavy rains, **lightning** is another damaging feature of thunderstorms. The amount of electricity in one bolt of lightning can supply a large city for a year, but we currently don't have a method for storing it. The temperature of lightning can reach as high as 28,000°C or about five times the temperature of the surface of the Sun. Lightning is formed when water droplets are pulled apart, creating positive (+) and negative (–) charges. The like charges gather together until the bolt of lightning is discharged. The discharge of electricity allows the system to be neutralized. Lightning can go from cloud to cloud or cloud to ground. The second scenario is the damaging type of lightning.

Thunder, the sound that occurs as the lightning strikes, is the compression of air. The sound travels about 1 mi in 5 seconds (1 km in 3 seconds), whereas the lightning is basically seen instantly. Heat lightning is cloud-to-cloud lightning or lightning that is too far away to be heard.

Precautions

Several steps can be taken to be safe during a thunderstorm. If you are near a building, it is best to get inside. While inside, stay away from televisions, corded phones, and the bathtub or shower. All of these can conduct electricity. If you are outside and cannot take cover indoors, stay low. Squatting in a field is better than lying down. In doing so, less area is in contact with the ground, which lessens the amount of grounding that could take place. Trees aren't the best place to take cover. Their height and water act as tall lightning rods. When a tree is hit by lightning, the sap boils and expands, causing the tree to explode.

Tornadoes

Tornadoes are narrow, funnel-shaped columns of wind that spiral around. Most spin in a counterclockwise direction, but some rotate in a clockwise motion. More tornadoes occur in the United States than anywhere else in the world. The clashing of cold, dry air from Canada and warm, moist air from the Gulf of Mexico fuel the thunderstorms and supercells that spawn twisters. The area around Kansas and Oklahoma is the site of so many tornadoes that it is called "Tornado Alley."

The air in a thunderstorm rotates around and starts to spin downward. Before this rotation reaches the ground, it is a funnel cloud. When it touches down, the tornado begins. The base of a tornado is usually less than ½ km across, but it moves back and forth, affecting a greater area. The winds in a tornado can reach up to 500 kmph (300 mph), while the twister itself moves at 40 to 65 kmph (25 to 40 mph). The **Fujita scale,** named for its developer, T. Theodore Fujita, measures the intensity of tornadoes based on wind speed and damage that occurs. The scale ranges from F0 (light damage, winds up to 72 mph) to F5 (extreme damage, winds over 261 mph). The wind speed can be determined by the type of damage that occurs or how far objects are impaled into other objects. If a tornado forms over water, it is a waterspout. It is rare for a waterspout to make landfall.

Hurricanes

A cluster of thunderstorms that stays together for a few days can begin to rotate. This occurs over warm ocean water (over 80°F). Water that is evaporated from the surface condenses in the cloud, releasing a tremendous amount of energy. This feeds the system even more, causing a low-pressure system to form. As the system grows, the winds at the center of it rise upward. If these winds are allowed to spiral outward at the top of the system, the system will grow and gain strength. A **tropical depression** is formed. When the maximum sustained winds reach 39 mph, it becomes a **tropical storm** and is named. If the maximum sustained winds reach 74 mph, it is called a **hurricane.** Once the storm reaches hurricane status, it is classified by the **Safir-Simpson Scale.** This rates the hurricanes based on central surface air pressure, maximum sustained wind speeds, storm surge, and damage (actual or potential). The scale ranges from Category I to Category V. These values can be seen in Table 16-3.

Table 16-3 Saffir / Simpson Hurricane Scale

Scale Number (Category)	*Pressure (millibars)*	*Winds (mph)*	*Storm Surge (ft)*	*Damage*
Trop. Depression	– –	<38	– –	– –
Tropical Storm	– –	39–73	– –	– –
I	>979	74–95	4–5	Minimal
II	965–979	96–110	6–8	Moderate
III	945–964	111–130	9–12	Extensive
IV	920–944	131–155	13–18	Extreme
V	<920	>155	>18	Catastrophic

The naming of hurricanes started in 1953, using female names. Prior to that time they were named for the area that they impacted the most. In 1979, male names were added, alternating with the female names. Each year the National Weather Service generates a list of names to be used in alphabetical order. The NWS currently has six lists of names which it uses in a six-year rotation. If a hurricane is particularly devastating, the name is retired.

The global winds act as steering currents for hurricanes. As the hurricane grows, an **eye** develops in the center. This is an area of calm winds and clear skies. Around the eye is the **eye wall,** which is a wall of clouds and strong winds. Around the eye and eye wall are spiraling bands of clouds and heavy rain. These feed the system with energy. Besides the sustained winds and torrential rains, hurricanes drag a bubble of water along with them. This **storm surge** can flood areas. As a hurricane makes landfall, a lot of friction with the ground occurs. This can result in tornado formation.

After the hurricane makes landfall, it loses its energy source by not being over the warm ocean waters. Within a few days it falls apart, usually just drenching areas with rain. The winds don't usually factor in damage at this point.

When warnings and watches are issued for hurricanes, several steps should be followed to ensure safety. Buildings and windows should be boarded up. All loose items should be put away or firmly secured. Water, canned food (with a manual can opener), and first-aid kits should gathered. Evacuations may be ordered by local authorities. A plan for meeting your family members should be in place for any emergency.

Winter Storms

Strong storms can also form in the winter. A blizzard is experienced when temperatures are low (below 20°F), winds are at least 35 mph, and it is snowing. The winds blow the snow around so that the visibility is below ¼ mile or less for at least three hours.

Synoptic Weather Maps

The characteristics of air are observed and recorded by **radiosondes.** They are weather balloons that transmit data back to the surface. These readings are combined for all weather stations and compiled by computers. By using this data, maps are generated for different levels of the troposphere and different areas of the world. Synoptic weather maps show the weather features at the surface such as pressure, temperature, precipitation, winds, fronts, and high- or low-pressure centers.

Station Models

The surface observations and weather data for a city are recorded on a **station model.** There are about 20 different parameters that can be recorded. These are compiled on a large map every three hours to create a surface weather map. Fronts, high-pressure systems, and low-pressure systems are also drawn in on the maps. Figure 16-8 shows a station model.

Figure 16-8 Weather station model and present weather symbols.

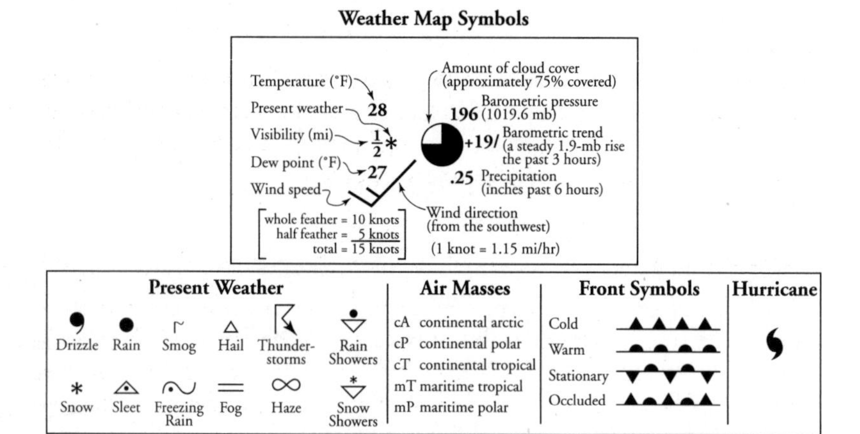

Temperature

Lines connecting equal temperature values are called **isotherms.** They are like contour lines for temperature readings. These lines shift more over continents, where there are greater fluctuations in temperature than over the oceans. They also move more in the Northern Hemisphere than in the Southern Hemisphere. This is due to more land existing above the equator and a higher percentage of water in the Southern Hemisphere. The isotherms also fluctuate across the United States depending on the season. The lines bend southward in the summertime in the center of the United States. The coastlines stay cooler than the center of the country at the same latitude due to the nearby oceans. This reverses in the winter, when the heartland of the United States is much cooler than the coasts and the isotherms bend upward. Figure 16-9 shows the isotherm variations.

Figure 16-9 General isotherm patterns for the United States.

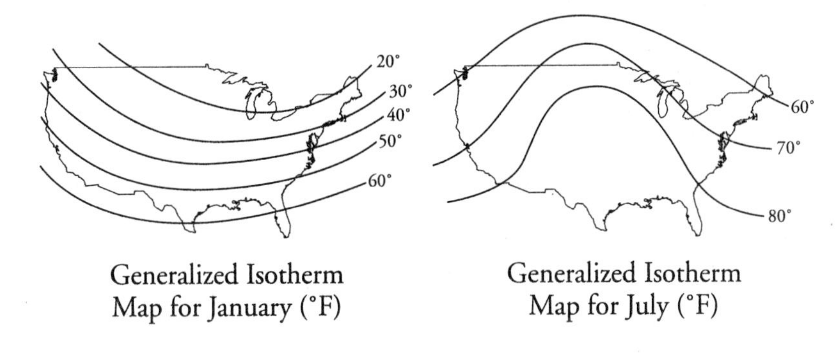

Pressure

The lines that connect areas of equal pressure are called **isobars.** The maps that show the isobars are useful in helping to show storms and prepare weather forecasts. They show areas of high pressure and low pressure. The surface winds usually go parallel to the isobars.

Forecasting

Some computers take all of the data from the radiosondes and run computer (mathematical) models to create forecasts. With the use of these readings, satellite photos, radar images, and information from aircraft, forecasts and predictions are made. These forecast maps are drawn two times per day (at 0700 and 1900 EST). The satellite images come from two different types of satellites. Geosynchronous (geostationary) satellites stay in orbit directly over one spot above the Earth. A circumpolar satellite orbits around the Earth in a path that takes it over the North and South poles. This satellite moves at a much faster pace, but can send images covering a larger area.

Chapter Checkout

Q&A

1. The characteristics of an air mass that formed over the Gulf of Mexico would probably be
   a. cool and dry.
   b. warm and dry.
   c. cool and humid.
   d. warm and humid.

2. What is the relative humidity of a sample of air that has a dry-bulb temperature of 20°C and a wet-bulb temperature of 11°C?

   a. 2 percent
   b. 9 percent
   c. 17 percent
   d. 30 percent

3. In order for clouds to form, cooling air must be

   a. saturated and have no condensation nuclei.
   b. saturated and have condensation nuclei.
   c. unsaturated and have no condensation nuclei.
   d. unsaturated and have condensation nuclei.

4. A low-pressure center located in the Midwestern United States generally moves toward the

   a. northeast.
   b. southeast.
   c. northwest.
   d. southwest.

Answers: 1. d **2.** d **3.** b **4.** a

Chapter 17
THE OCEANS

Chapter Checkin

- ❑ Determining the differences between underwater features
- ❑ Understanding the composition of water
- ❑ Knowing the surface currents of the oceans

Approximately 70 percent of the surface of the Earth is covered by water, with most of that water being the oceans. While the highest point of land reaches up 8.8 km (29,035 ft), Mt. Everest is no match for the depth of the **Marianas Trench.** This trench is found off the coast of Japan and reaches down 11 km below sea level.

Salinity

Salts are part of the sediments from the mountains and land that rivers wash into the ocean. The sediments settle to the bottom, forming new sedimentary rocks. The water evaporates and leaves the salts behind, which are dissolved in the ocean water. Over time, the oceans are getting saltier. The average salinity for the oceans is about 35 parts per thousand (ppt). This means that for every 1,000 g of water, 35 gm are salt and 965 g are water. The Red Sea is very salty at 40 ppt. Areas where rivers empty into the ocean are less saline. In the tropics where the sun is strong, the water is saltier due to increased evaporation.

The **salinity** of water is found by using a meter that tests for electrical conductivity. The ocean salt is composed of seven major salts, including NaCl, MgCl, KCl, and $CaCl_2$. More than 55 elements have been found in seawater.

Layers of the Ocean

The ocean has several layers to it. The surface layer is called the mixed layer. This is where the rivers flow into the ocean while water evaporates into the air. This layer is warmed by the Sun and is mixed by the wind and waves. The mixed layer reaches down 100 to 300 m. At the bottom of the mixed layer is a **thermocline,** which is a zone where the temperature changes rapidly. This zone goes down to about 1,000 m. Thermoclines also exist in lakes. If you dive down deeply enough, you can feel the water temperature drop rapidly. Below the thermocline is the deep-water zone. This is a stable area of cold water and is rich in nutrients. The increased nutrient concentration is due to the decomposition of dead plants and animals that occur at the bottom. Areas of **upwelling** pull water up from this area, creating rich fishing areas.

Surface Currents

The water on the surface of the ocean is affected by several factors. Winds blow the water while the Moon's gravity pulls on it. The water moves, but is also affected by land masses, which get in the way of the flow. As this occurs, the Earth rotates, adding the Coriolis effect. The result is a complex system of oceanic circulation. In general, the oceans move clockwise in the Northern Hemisphere and counterclockwise in the Southern Hemisphere. Water in the tropics warms up and moves north, while cold, polar waters move south. The effects on climate are discussed in Chapter 18. Figure 17-1 shows the global oceanic circulations.

Figure 17-1 Surface ocean currents.

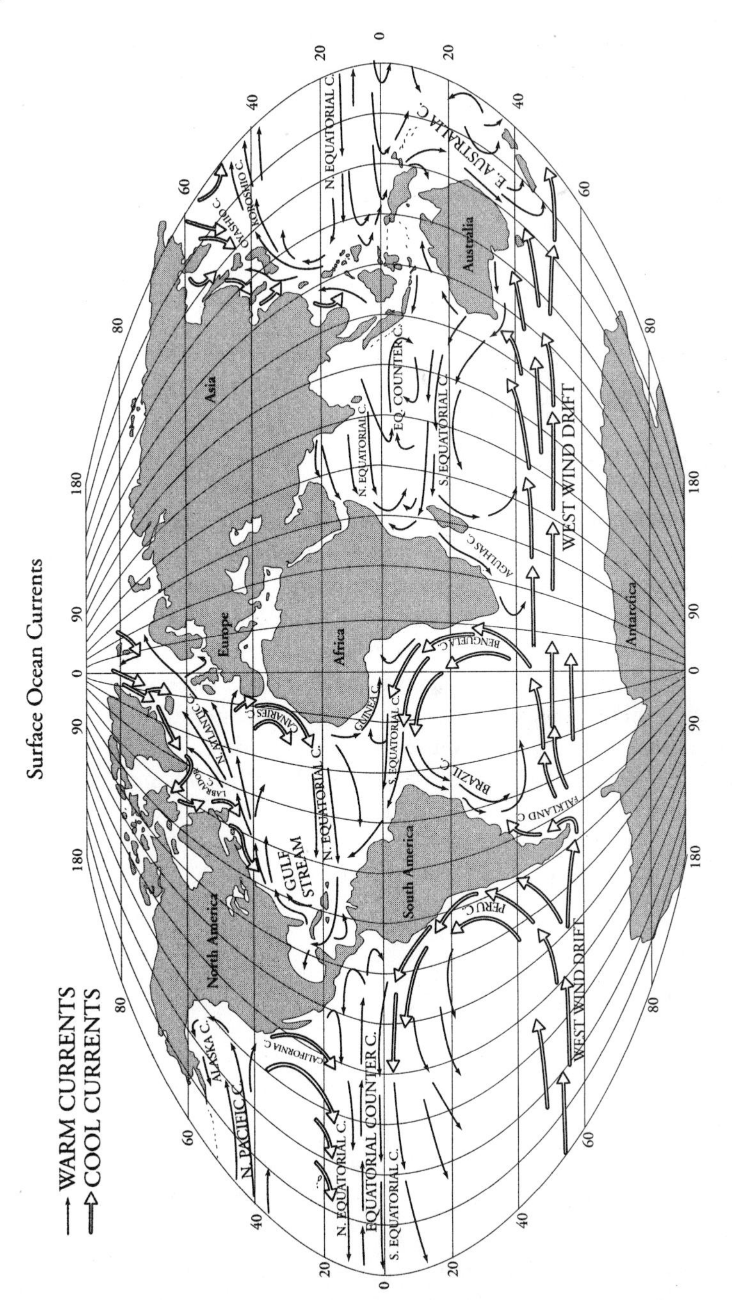

Life in the Ocean

Phytoplankton and algae are abundant in the mixed layer. Phytoplankton are microscopic plants that float in the water and move to wherever the currents take them. Algae are macroscopic plants that can float in the water or become attached to a solid object. The Sun is the ultimate source for their life, and the rest of the food chain is dependent on their survival. Diatoms are one type of phytoplankton that have shells made of silica. The silica is important in helping cement sediments together when sedimentary rocks are formed. Zooplankton are microscopic animals that feed on phytoplankton. These animals are eaten by anything from shrimp and small fish to baleen whales.

Ocean Floor Features

Soundings taken from ships show that the ocean floor isn't flat. Coming off of the edge of the continents is the **continental shelf.** This gradually slopes downward until it reaches the **continental slope.** At this point, the slope is relatively steep until it reaches close to the ocean floor. As the slope gets near to the floor, the gradient gets lower. This is the **continental rise. Turbidity currents** (undersea mudslides) are found here. The ocean floor is comparatively level, aside from other features found there. **Seamounts** are underwater mountains and are usually volcanic in origin. Flat-topped seamounts are **guyots.** The tops were leveled by ocean waves when they were above sea level. Over time, **subsidence** caused the guyots to sink below sea level. Fringe reefs surround islands in warmer climates. If the islands sink down, these coral reefs can survive by growing on top of each other, creating an **atoll.** Deep in the ocean are **trenches.** These form where plates converge and a subduction zone is present. These features can be seen in Figure 17-2.

Figure 17-2 Ocean bottom features.

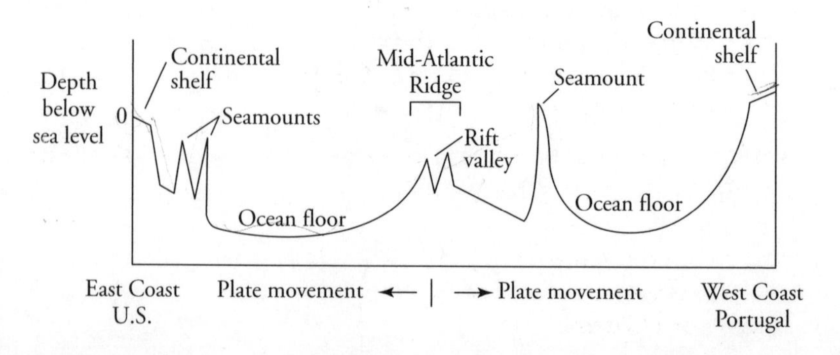

Fracture Zones

Along the rift valleys, between diverging plates are areas where containing transform (strike-slip) faults. The rift valley and associated ridge are divided up into separate pieces called fracture zones. The rift valley is not a continuous line running between the plates, but a series of smaller sections that are offset to each other. Movement along these faults can create earthquakes. The locations of these offset pieces can be seen in Chapter 10, Figure 10.2.

Hydrothermal Vents

When the plates are diverging, a mid-ocean ridge and rift are formed. Hydrothermal vents have been discovered along these ridges and rifts, including entirely self-contained ecosystems. Instead of utilizing the process of photosynthesis, organisms of the bottom of the food chain use chemosynthesis. Because light cannot penetrate to the bottom of the ocean, sulfur is abundant and is used instead. The sulfur comes from lava that has come through the Earth's crust, generally along rift areas. Giant clams and huge crabs live here near these hydrothermal vents, along with a multitude of strange and mostly unidentified fishes. Some vents have a hot spring of hydrogen sulfide that pours out of the ocean bottom in a black plume, like a chimney. The black color is caused by a chemical reaction that occurs as the ocean water mixes with water coming out. Iron sulfide is produced in this reaction, which is black in color. These black smokers also help supply the sulfur needs of the producers in this isolated food chain. These have been observed in several areas and was first discovered along the East Pacific Rise.

Chapter Checkout

Q&A

1. Most of the Earth's surface ocean currents are caused by
   - a. stream flow from continents.
   - b. differences in ocean-water density.
   - c. the revolution of the Earth.
   - d. the prevailing winds.

2. Which ocean current cools the climate of some locations along the western coastline of North America?
   - a. Florida Current
   - b. California Current
   - c. Canaries Current
   - d. Alaska Current

3. Which of the following aren't found on the ocean floor?

a. Guyots
b. Trenches
c. Seamounts
d. Glaciers

4. Surface currents are affected by

a. Tilt of Earth's axis.
b. Earth's revolution.
c. Coriolis forces.
d. Adiabatic lapse rates.

Answers: 1. d **2.** b **3.** d **4.** c

Chapter 18
WATER AND CLIMATE

Chapter Checkin

- ❑ Knowing how land and water heat at different rates
- ❑ Determining how surface features and ocean currents alter climate
- ❑ Understanding the factors that affect the climate for a region

Water is the one of the key ingredients in determining the climate of a region. Temperature is the other main component. Many other factors are added in to establish the characteristic weather patterns for a region.

The Water Cycle

The water cycle is the endless movement of water through the Earth and atmosphere. Water is put into the air by transpiration by plants and evaporation. This water vapor condenses and forms clouds and precipitation. The precipitation falls to Earth and sinks into the ground or forms runoff, where the cycle starts again.

Climate

Climate is the overall weather conditions for an area. Looking at the yearly climate data for a region doesn't tell you the whole story, however. This is very evident in areas that have monsoons. Although part of the year is very rainy, other parts are much drier. The two main factors that determine the climate are temperature and rainfall. Other variables that are taken into consideration are the amount of sunlight (number of hours and days), the wind (direction, speed, and steadiness), and pollution.

Factors Affecting Climate

The local climate for a region depends on several factors. The primary six factors are latitude, altitude, prevailing winds, topography, distance from large bodies of water, and nearby ocean currents. Temperature and precipitation are affected by changes in the shape of the land and surface features.

Factors Controlling Temperature

The latitude for an area can affect the temperature for that region. Along the equatorial areas it is warm, with little variation in temperature. In the mid-latitude areas (40° to 45° N and S), there are longer days in the summer. This leads to larger temperature ranges. The polar regions experience dark, cold winters and summers of low-angle sun. The altitude of a region can affect the temperatures. As the altitude increases, the temperature decreases.

The prevailing winds for an area create local temperature variations. Along the west coast of the United States, the winds coming off the cool Pacific Ocean create cooler summers and mild winters. The east coast of the United States experiences the reverse effect. The winds coming from the land cause the summers to be hotter and the winters to be colder. A range of mountains can block these winds from altering the temperatures.

The distance from a body of water can affect the yearly temperature ranges. A continental climate has a larger yearly temperature range than a coastal climate. Areas near an ocean (or large lake) have a moderated climate with smaller yearly temperature ranges. This is due to the ability of water to retain heat. The cold and warm ocean currents affect the temperature ranges for an area. The Gulf Stream keeps the British Isles warmer than other areas with the same latitude.

Factors Controlling Precipitation

Global wind patterns affect the rainfall for an area. The converging global winds along the equator rise up away from the surface creating an area of constant convection. This region, known as the Intertropical Convergence Zone (ITCZ), is an area containing many thunderstorms. This region migrates north and south of the Equator with the vertical ray of the Sun. Along the ITCZ, the equatorial regions, and 60° (N and S) latitude, precipitation amounts are greater due to the converging air that rises upward. Where the air is sinking and diverging at the surface (30° and 90° N and S), dry conditions are experienced. Deserts and tundras are found at these latitudes. Mountains have wet and dry sides. This is explained later in this chapter. The warm winds that come down off of the east side of the Rocky Mountains are Chinook winds. Santa Ana winds are warm, dry winds that come down the west side of the Rockies and affect southern California. Areas that are near warm-water currents, such as the east coast of the United States, receive more rainfall. Areas near a body of water tend to experience more fog than areas farther away.

Altitude

As noted in Chapter 16, the temperature decreases with increasing altitude in the troposphere. The normal rate of this decrease is about 1°C/160 m (5°F/1,000 ft). When air rises, this rate changes to 1°C/100 m due to the changes in pressure and relative humidity. Cold air in a valley region can be trapped when warmer air comes on top of it. This is called an **inversion.** This creates an area where the air quality can be poor. The lack of air circulation keeps the pollution from the region trapped in the valley.

Unequal Heating Rates of Water and Land

Water is the most difficult natural material on Earth to heat up. It also takes the longest to cool down. This property of water can affect the climate of an area. Table 18-1 shows the specific heats of natural materials that are commonly found on the Earth.

Table 18-1 Specific Heats of Common Materials

MATERIAL	SPECIFIC HEAT (calories/gram • C°)
Water – solid	0.5
Water – liquid	1.0
Water – gas	0.5
Dry air	0.24
Basalt	0.20
Granite	0.19
Iron	0.11
Copper	0.09
Lead	0.03

Several other factors affect the heating of a large body of water. Besides having a high specific heat, water is **transparent.** Light rays that reach a body of water penetrate more deeply. This spreads out the heat that is absorbed into a larger volume. Some of the energy absorbed by the body of water is used in the evaporation process. For these reasons, water heats up and cools off more slowly than land. This affects the local climate of an area that has a large body of water nearby.

Land can have many varied surfaces, which affect its ability to absorb and reradiate energy. The land areas can be lighter or darker in color. The darker areas absorb energy better than lighter colors, which reflect the light better. If an object absorbs energy quickly, it cools off quickly as well. The roughness or smoothness of a surface also affects how much insolation is absorbed. A rough surface absorbs more light as opposed to a smooth surface, which reflects more light.

The amount of moisture in the land also affects how quickly it will heat up and cool down. Drier land areas heat up and cool down more quickly than moister areas. The amount of grass in comparison to the amount of pavement in some areas can affect the temperature of an area. Pavement heats up much more quickly than grassy areas. This is one of the reasons for the **urban island effect.** This is a situation where the city is warmer than the surrounding areas. The heat absorbed and generated by buildings and vehicles adds to the total heat of the area. In some areas where there is snow and ice, the amount of insolation absorbed is less. Much of the sunlight is reflected due to the smooth, white surface (which is also shiny where there is ice). The reflection of sunlight by snow and ice is referred to as **albedo.**

Mountains

As air moves across a mountainous region, it is lifted upward. This expanding parcel of air cools, which causes clouds to form and precipitation to occur. The moisture is squeezed out of the air like a sponge filled with water. The air moves over the mountain and downward into the valley. The air warms up due to being compressed. It actually heats up quickly because it has less moisture than before it precipitated out on the other side of the mountain. The result on the land is the formation of a desert or area with very dry conditions. The **windward** side of the mountain is more rainy and cloudy than the **leeward** side. This can be seen in Figure 18-1.

Figure 18-1 Precipitation variations on either side of a mountain.

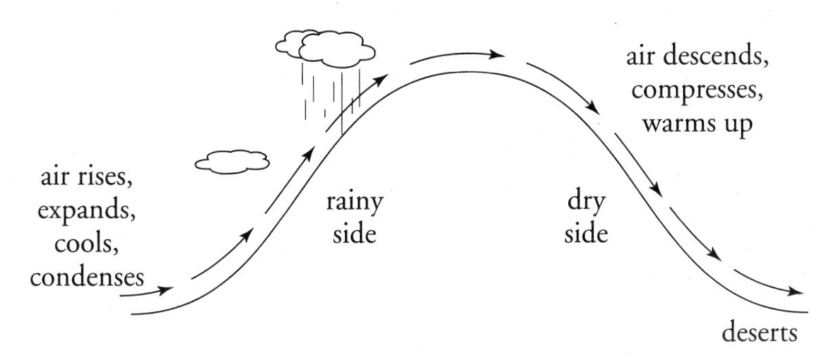

Long-Term Climate Change

The Earth's energy budget is currently in balance. Climate change can be viewed in two ways: short term and long term. Short-term effects last for less than 20 years. Factors that affect this are volcanic eruptions. The ash and gases that are put into the air can block sunlight from reaching the Earth or trap the heat that is already here. **El Niño** and **La Niña** episodes affect global weather patterns. These events are connected to the prevailing winds along the equatorial regions. The warm water is pushed to the western side of the oceans. When the winds lessen, the bulge of warm water moves eastward. An El Niño event then begins, which disrupts ocean currents and weather patterns. The reverse of this is La Niña. There is no pattern for these occurrences.

Long-term effects on the climate are periods of glaciation and interglacial periods. The last glacier ended about 15,000 years ago. These ice ages occur about every 100,000 years, but this isn't a regular pattern throughout the history of the Earth. The wobble of the Earth's rotation and revolution affect the worldwide weather and climate patterns. The time periods related to this wobble are called **Milankovic periods.** These periods take into account the eccentricity, axial tilt, and precession of the Earth's orbit. The combination of the variance of these patterns results in a 100,000-year ice-age pattern and affect weather patterns on a global level.

Chapter Checkout

Q&A

1. How does the average annual surface temperature compare from latitude to latitude?

    a. As latitude increases, the average annual surface temperature decreases.
    b. As latitude increases, the average annual surface temperature increases.
    c. As latitude increases, the average annual surface temperature remains the same.
    d. As latitude increases, the average annual surface temperature rises and falls with the tides.

2. Compared to a coastal location of the same elevation and latitude, an inland location is likely to have

    a. warmer summers and cooler winters.
    b. warmer summers and warmer winters.
    c. cooler summers and cooler winters.
    d. cooler summers and warmer winters.

3. At which latitudes do currents of dry, sinking air cause the dry conditions of Earth's major deserts?

    a. 30° N and 30° S
    b. 60° S and 90° S
    c. 0° and 30° N
    d. 60° N and 60° S

4. Compared to land surface temperature changes, water surface temperature changes occur

    a. more slowly because water has a lower specific heat.
    b. more slowly because water has a higher specific heat.
    c. faster because water has a lower specific heat.
    d. faster because water has a higher specific heat.

Answers: 1. a **2.** a **3.** a **4.** b

Chapter 19
THE UNIVERSE

Chapter Checkin

- ❑ Understanding the big-bang hypothesis
- ❑ Knowing the members of the solar system
- ❑ Learning about the electromagnetic spectrum

The objects outside of Earth's atmosphere comprise the solar system and the universe. Our solar system is comprised of planets, the Sun, moons, asteroids, meteoroids, and comets. Our solar system is a part of the Milky Way Galaxy. Many other stars and solar systems lie in our galaxy. Beyond the Milky Way are many other galaxies.

Origin of the Universe: The Big-Bang Hypothesis

About 15 billion years ago, the entire universe was compacted into a single point called a singularity. The sphere exploded in all directions, and a giant cloud was formed. Some parts of the cloud moved faster than others, and many parts condensed into galaxies. Billions of these galaxies were formed. Within these galaxies, dust and gases spiraled around, collecting other pieces of gas and dust. These eventually became stars and planets. This is the theory of the origin of the universe known as the **big-bang hypothesis.** According to this theory, the universe is still expanding. The observations of distant galaxies support this idea. Some light from stars that is reaching the Earth now has traveled a very long distance. In fact, some of these stars no longer exist.

Galaxies and Stars

A galaxy is comprised of billions of stars kept together by gravity. Some galaxies have more than 100 billion stars. Some estimates put the number of galaxies in the universe at more than 100 billion.

The shape of a galaxy can fall into one of several categories. Our solar system belongs to the **Milky Way galaxy,** which is a **spiral galaxy** and consists of more than 200 billion stars. Spiral galaxies have bands of stars that revolve around a central point. Other galaxy shapes include elliptical and irregular. Elliptical galaxies are shaped like a football or a three-dimensional oval. Irregular galaxies are those that don't fit into the spiral or elliptical categories. The nearest galaxy to us is the Andromeda Galaxy at about 2–3 million light years away.

On a clear night, hundreds of stars or more can be seen with the unaided eye. In areas affected by light pollution (street lights, city lights) the amount seen decreases. Many more can be seen with telescopes. These stars can be viewed as patterns in the sky called constellations, which can also be used as navigational aids.

Telescopes

Several different types of telescopes are used to study stars, planets, and other objects in the solar system. The refracting telescope has two lenses and shows an upright image. A reflecting telescope consists of a light-gathering tube with a curved mirror at the end. The condensed light is magnified through a lens. Reflecting telescopes are used to see dim and faraway objects. The structure of these telescopes is shown in Figure 19-1.

Figure 19-1 Structure of refracting and reflecting telescopes.

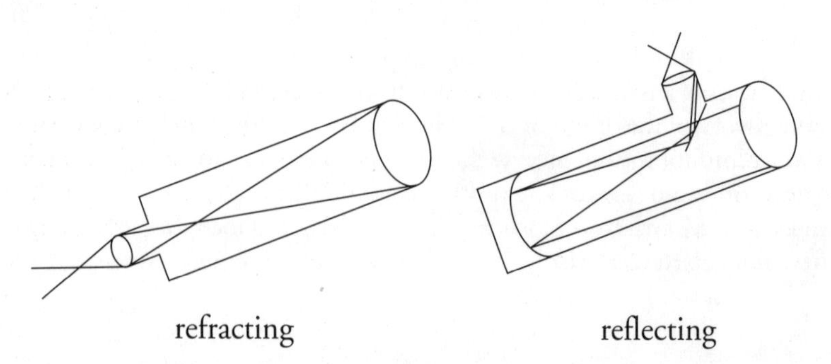

The last type of telescope gathers light and radio waves and is called a radio telescope. A computer changes the information gathered by a radio telescope and transforms it into a picture that can be interpreted.

Electromagnetic Spectrum

The **electromagnetic spectrum** is a diagram showing the wavelengths of energy divided into different categories. The high-energy bands have very short wavelengths. As the energy level decreases, the wavelength increases. All of these waves move at the same speed. The different groups of waves can be seen in Figure 19-2.

Figure 19-2 The electromagnetic spectrum.

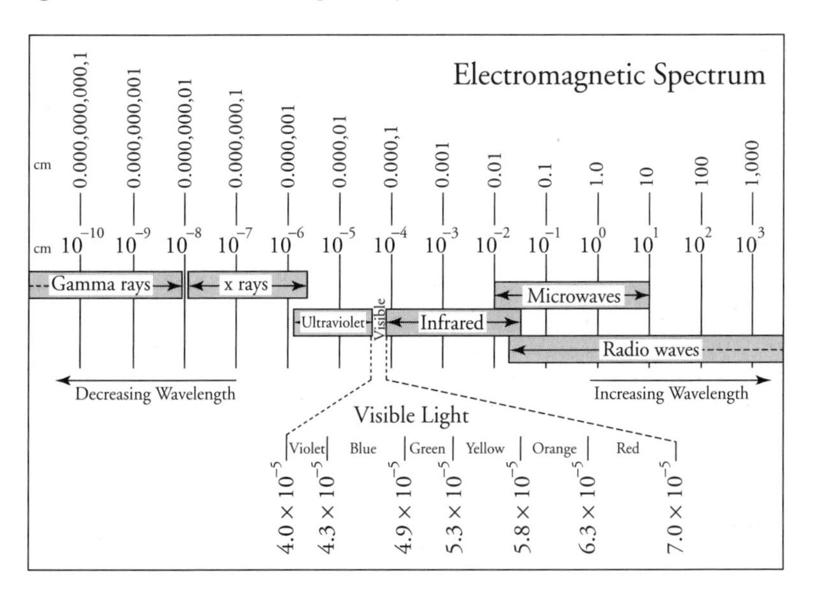

Visible light consists of a small band in the electromagnetic spectrum. A **spectroscope** separates incoming visible light into its component parts. The colors seen in a spectroscope are the wavelengths of light emitted (given off) by a star or other radiant body. This is called the bright-line spectrum. The dark areas between these bands are the wavelengths of light that are absorbed by the body. These are the dark-line spectrum. If the light seen through a spectroscope has no unbroken bands of light, it is called a continuous spectrum. The bright-line spectrum and the dark-line spectrum are used to identify what materials stars are made of. Each element has a specific set of spectra, making it unique. By comparing these to what is observed in the emission spectrum of the star, the elements present in the star can be determined.

Doppler Shift

As an object moves away from you or moves toward you, there is a small change in the wavelengths of light that are given off. If the object is coming toward you, the waves are compressed, making the wavelength shorter. This moves the bands of light to the blue (or shorter) side of the visible light spectrum and is known as a **blue shift.** If the object is moving away from you, the waves are stretched out, making them longer. This moves the band of light seen toward the red (or longer) side of the visible light range and is called a **red shift.** This has been helpful in determining the origin of the universe. The red shift/blue shift phenomenon is similar to listening to a car race. As the car approaches you, the pitch of the car's sound gets higher. As the car passes you, there is a long, droning, lower-wavelength sound. During this time, the driver of the car hears a constant sound coming from the engine because he or she is moving with the car.

Constellations

When stars are viewed as a group or a pattern, they are called a **constellation.** There are currently 88 named constellations. We do not see all of them at any one time or over the period of a year. On a daily basis, they appear to rise in the eastern sky. They move through the sky until they appear to set in the west. This is because the Earth is rotating. Some constellations are seen only during certain seasons or periods of time. This is due to the revolution of the Earth around the Sun. At some points in the year, the Sun is blocking some constellations with light. The constellations of the zodiac (such as Virgo and Taurus) are a better-known group of these seasonal constellations. Orion is another seasonal constellation, which appears only in the winter months for observers in the United States. Figure 19-3 shows the positions of the Earth, Sun, and zodiac constellations. The zodiac sign coincides with the position of the Sun.

Figure 19-3 Positions of the Earth, Sun, and zodiac constellations.

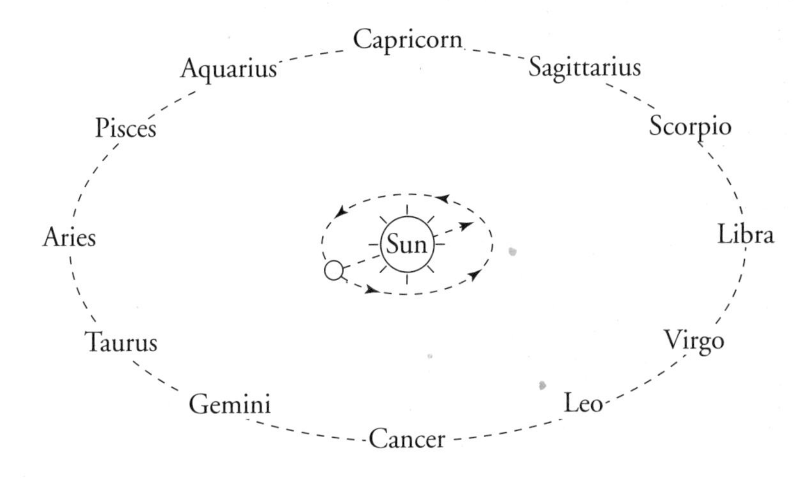

Some constellations are seen all year, such as the Big Dipper and the Little Dipper—but only in the Northern Hemisphere. There are many more constellations that people who live in the Northern Hemisphere will never see. These constellations, such as the Southern Cross, are seen only in the Southern Hemisphere. Some constellations have subsets or supersets of stars that are known by themselves. These are asterisms. Some of the more well known ones are the Big Dipper (part of the Great Bear), Pleiades (part of Taurus) and the Belt of Orion—known in Latin America as The Three Marys (part of Orion).

Distances to Stars

Several units are used to measure the distance to objects in the solar system and beyond. Miles or kilometers could be used, but the size of the numbers gets so large that errors could be made. Zeros or decimal places could easily be lost or added, making the number inaccurate. **Astronomical units** are used to show distances within the solar system. This unit is the average distance between the Earth and the Sun, which is about 150,000,000 km or about 93,000,000 miles. Beyond our solar system, the **light year** is used as the convenient unit of measurement. As the name implies, it is the distance that light travels in one year. Because light travels about 299,792 km/sec (186,282 mi/sec), this amounts to about 9.46 trillion km/yr (5.88

trillion mi/year). To get an understanding of how far this is in other terms, a jet plane going 500 mph (800 kph) would take 1.34 million years to go 1 light year. On a smaller scale, light travels about 7.48 times around the Earth in 1 second!

Star Properties

Several properties help to identify a star. A star is basically a large ball of burning gas that is held together by gravity. The color of a star depends on its temperature. Cooler stars are reddish in color. As the temperature increases, the color changes to yellow, then white, and then blue. The types and amounts of elements present in a star make it unique. Most stars have a basic makeup of hydrogen and helium, but many other elements can be present. These elements come from the fusing of hydrogen and helium atoms. As the age and size of the size of a star increases, then the likelihood of other elements increases as well.

The brightness of a star can be interpreted in several ways. The **apparent magnitude** is how bright a star looks to us here on Earth. For example, a star could actually be very bright, but if it is very far from Earth, it will seem dimmer than a star that is less bright but closer to our planet. The **luminosity** is the actual brightness of a star. The **absolute magnitude** refers to how bright a star looks from a specific distance, 32.6 light years away, to make it easier to compare. Stars can vary in how they shine. Most shine constantly and emit a steady amount of light. Other stars are variable in the light they put out. These are **Cepheid** stars. **Pulsars** are stars that send out their energy in pulses.

The size of the star can vary as well. They can range from being as small as a white dwarf (smaller than the Earth), to as large as a red giant (10 to 100 times the size of our Sun), to a supergiant (100 to 1,000 times the size of the Sun).

Life Cycle of Stars

Clouds of gas collect together in space. If this **nebula** contracts due to gravity and begins to burn, a star is then formed. This star can burn for billions of years before its life cycle ends. As the star nears the end of life, it will expand and become a **red giant.** Our Sun will probably expand out to about the orbit of Mars. Its life will end as it explodes, becoming a **supernova.** Or it will collapse onto itself, becoming a **white dwarf.** Our Sun will probably go the route to become a white dwarf. The Sun is about halfway through its life cycle. This means that the expansion and collapse will not happen for about another 4.5 billion years.

The Hertzsprung-Russell diagram plots stars based on luminosity and temperature. Stars that fall into a broad band on the H-R diagram are called **main sequence** stars. This is how a star spends most of its life. Figure 19-4 shows the H-R diagram.

Figure 19-4 Luminosity and temperature of stars (the H-R diagram). (Name in italics refers to star shown by a ⊕.)

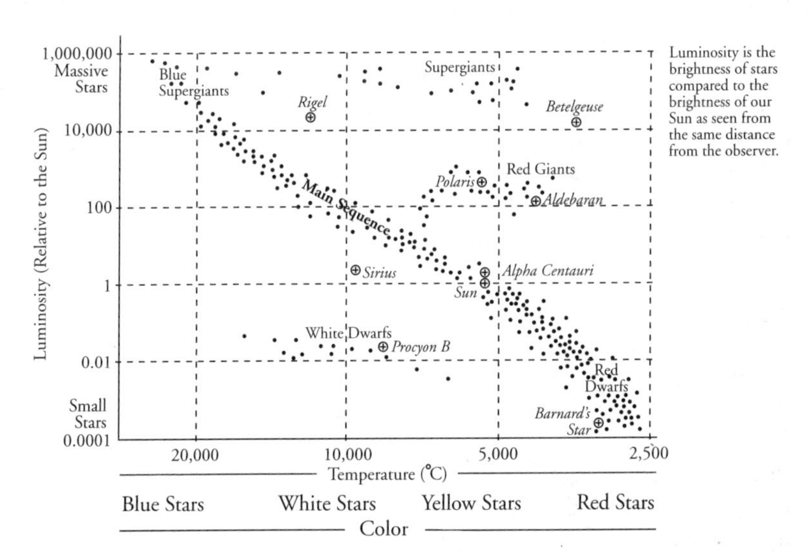

The Sun

Our Sun is extremely important to us. Without its light and heat, life on Earth wouldn't exist. When its size and temperature are compared to other stars, it is a very average star. It burns hydrogen gas that fuses and becomes helium. Light takes about 8 minutes and 20 seconds to reach the Earth. The Sun rotates around once in about 27 days while also revolving around the Milky Way galaxy once every 220 million years. Galileo was one of the first Western scientists to notice that the Sun rotated. He knew this by using his telescope to observe sunspots over time. The exterior temperature of the Sun is about 5,500°C, whereas the interior is estimated to be about 15,000,000°C. Its diameter is about 110 times that of the Earth, and it's volume could hold more than 1 million Earths.

Structure of the Sun

The outer yellow layer of the Sun's surface is the **photosphere** and is about 400 km thick. This is made up of individual cells called **granules** that are about 1,500 km in diameter. The next layer out is the **chromosphere,** which is a reddish area of glowing hydrogen. Surrounding these is the **corona,** which is usually seen in a solar eclipse. Extending out from the corona are **solar prominences.** These prominences are huge red flamelike arcs that reach outward and come back to the corona.

Sunspots

Sunspots are cooler areas in the Sun, although if they were taken off the surface of the sun they would still glow. They simply appear darker since they are cooler by a few thousand degrees and represent an area where the magnetic field of the sun is in a state of flux. Chinese astronomers have been observing sunspots for thousands of years by looking at the Sun during sunrise and sunset. The numbers of these sunspots present vary yearly and have about an 11-year cycle where the magnetic field of the sun reverses (over the span of 22 years). This means that the maximum number of sunspots present occurs about every 11 years. During years of increased solar activity, these flares can reach outward from the Sun, causing solar winds or streams of charged particles to be hurled into space. When these particles reach the Earth, they interact with the magnetic and electronic fields. Radio interference increases and **auroras** are created around the North and South poles. The Aurora Borealis ("northern lights") and Aurora Australis ("southern lights") can be seen in the higher latitudes and appear as different colors of light silently dancing in the sky.

Planets

The current model of the solar system states that it has nine planets. Recent discoveries are challenging this, adding to the total. If you observe the sky over a long period of time, the position of the stars is fixed. The planets change their positions, generally moving toward the east. The word *planet* is Greek for "wanderer." On occasion a planet may migrate toward the west, which is known as **retrograde motion** and has to do with our position in our orbit compared to that planets position, like two athletes running around in different lanes on a track.

Other members of the solar system include many moons, asteroids, comets, and meteoroids. Data for each of the planets, the Moon, and the Sun can be seen in Figure 19-5.

Figure 19-5 Solar system data.

Object	Mean Distance from Sun (millions of km)	Period of Revolution	Period of Rotation	Eccentricity of Orbit	Equatorial Diameter (km)	Mass (Earth = 1)	Density (g/cm^3)	Number of Moons
SUN	—	—	27 days	—	1,392,000	333,000.00	1.4	—
MERCURY	57.9	88 days	59 days	0.206	4,880	0.553	5.4	0
VENUS	108.2	224.7 days	243 days	0.007	12,104	0.815	5.2	0
EARTH	149.6	365.26 days	23 hr 56 min 4 sec	0.017	12,756	1.00	5.5	1
MARS	227.9	687 days	24 hr 37 min 23 sec	0.093	6,787	0.1074	3.9	2
JUPITER	778.3	11.86 years	9 hr 37 min 23 sec	0.048	142,800	317.896	1.3	16
SATURN	1,427	29.46 years	10 hr 14 min	0.056	120,000	95.185	0.7	18
URANUS	2,869	84.0 years	17 hr 14 min	0.047	51,800	14.537	1.2	21
NEPTUNE	4,496	164.8 years	16 hr	0.009	49,500	17.151	1.7	8
PLUTO	5,900	247.7 years	6 days 9 hr	0.250	2,300	0.0025	2.0	1
EARTH'S MOON	149.6 (0.386 from Earth)	27.3 years	27 days 8 hr	0.055	3,476	0.0123	3.3	—

Inner Planets

Mercury, Venus, Earth, and Mars are known as the **terrestrial planets.** They all have rocky surfaces. Mercury is covered with impact craters and has a temperature range from 400°C during the day to –200°C at night. Venus is covered by clouds and has an atmosphere of mostly carbon dioxide. Small amounts of nitrogen are also found in the atmosphere, with some droplets of sulfuric acid. This leads to a severe **greenhouse effect** and causes Venus' extremely high surface temperatures of 460°C.

Because Mercury and Venus are closer to the Sun than the Earth, you see them around sunrise and sunset. They are called the morning star and the evening star because of this. The Earth has a nitrogen-based atmosphere, which supports life as we know it. It also has water in all three physical states: solid, liquid, and gas. Mars has polar ice caps made of frozen carbon dioxide (and one that is a mix of carbon dioxide and water) that change size seasonally. Mars has several extinct volcanoes, including the largest known volcano in our solar system, Olympus Mons, and also has one of the largest canyons in the solar system called the Valles Marineris (which would cross the United States if it was on Earth's surface!). Recent landings on Mars have added data to help support the idea that water once existed on the surface.

Outer Planets

The outer planets include the gas giants Jupiter, Saturn, Uranus, and Neptune. These **Jovian planets** all have a rocky core surrounded by a liquid mantle of mostly hydrogen. A layer of gaseous hydrogen and helium at the surface surrounds these layers. Each of the Jovian planets has a ring system and many moons.

Pluto, the smallest planet, is usually the furthest out from the Sun. Due to the highly elliptical orbit of Pluto and the nearly circular orbit of Neptune, however, they sometimes switch places in their order out from the Sun. This switch occurs for only a few years. The last time Pluto was closer was from 1979 to 1999. Since this switch occurs every 248 years, the next time it happens it will be the year 2227.

The Moon

The nearest object to Earth in the solar system is the Moon. The race to land on the Moon came from the Cold War between the United States and the USSR in the 1950s. President Kennedy made it a national goal to put a man on the Moon and have him return safely before 1970. Although he never saw this accomplishment, it occurred on July 20, 1969, when Apollo 11 landed in the Sea of Tranquility. The USSR actually beat the United States to orbiting the Earth and landing unmanned spacecraft on the Moon, but still hasn't put a human on the Moon.

Properties of the Moon

The side of the Moon that faces the Earth doesn't have as many **craters** as the side facing away. This is because the Moon's period of rotation is just about the same as its revolution around the Earth. Because of this tidal lock, the same side always faces outward, and it is constantly in the line of fire of meteorites. The lack of an atmosphere and weather on the Moon prevents these craters from being worn away, as is the case on Earth. The side facing the Earth has many dark, flat areas that were seen by Galileo. He called them **maria,** which means "seas" in Italian. They are not seas as we know them, but were probably areas of volcanic activity when the Moon was formed.

History of the Moon

Although the Moon's origin isn't exactly known, one theory has wide acceptance. It states that when the Earth was forming, another object collided with it. This object broke through the recently formed crust. Molten rock from below splattered out and cooled, forming the Moon. As the Moon cooled, it was showered with rock fragments, causing lots of craters and basins to form. The next stage in the development of the Moon was that lava flowed out from inside of the Moon into the basins. This went on for a billion years. The last 3 billion years have been relatively quiet in terms of volcanic eruptions and moonquakes. Meteoroids and mircometeoroids still pummel the Moon's surface.

Comets, Asteroids, and Meteoroids

Several other objects are in the solar system. A **comet** can be thought of as a dirty snowball made of ice, rock, and various elements (carbon, nitrogen, oxygen, sulfur, and magnesium). A comet moves in a elliptical orbit around the Sun. As the comet approaches the Sun, some of the comet's nucleus melts. This is blown away by solar winds and motion of the comet. This becomes the comet's tail.

Between Mars and Jupiter is the **asteriod belt.** This is a band of rocks that orbit the Sun. The origin of the asteroids is unknown. One theory says that these fragments are leftover from when the solar system was made. They possibly could have been another planet that never formed. Another theory states that they may be comets that are extinct or inactive. At times, a piece can come loose and head toward a neighboring planet.

A **meteoroid** is a piece of rock that is orbiting in the solar system. As meteoroids enter the Earth's atmosphere, they are called **meteors** or "shooting stars." Shooting stars are actually small pieces of rock that burn up as they enter the atmosphere. At different times of the year, meteor showers can be seen. They are named for the constellation that they appear to originate from. The better-known meteor showers are the Perseids in August, the Orionids in October, the Taurids in November, and the Geminids in December. All but the Orionids occur around the 10th of the month; the Orionids occur around the 20th. If the rocks reach the ground, they are called **meteorites.** Impact craters of various sizes have been found on the Earth. The weathering forces on the Earth wear these away, unlike craters on the Moon or other planets.

Chapter Checkout

Q&A

1. Which three planets are known as gas giants because of their large size and low density?
   - **a.** Venus, Neptune, and Jupiter
   - **b.** Jupiter, Saturn, and Venus
   - **c.** Jupiter, Saturn, and Uranus
   - **d.** Venus, Uranus, and Jupiter
2. Which planet has the most eccentric orbit?
   - **a.** Mercury
   - **b.** Venus
   - **c.** Neptune
   - **d.** Pluto
3. Which statement about electromagnetic energy is correct?
   - **a.** Violet light has a longer wavelength than red light.
   - **b.** X-rays have a longer wavelength than infrared waves.
   - **c.** Radio waves have a shorter wavelength than ultraviolet rays.
   - **d.** Gamma rays have a shorter wavelength than visible light.
4. According to the big bang theory, the universe began as an explosion and is still expanding. The theory is supported by observations that the stellar spectra of distant galaxies show a
   - **a.** concentration in the yellow portion of the spectrum.
   - **b.** concentration in the green portion of the spectrum.
   - **c.** shift toward the blue end of the spectrum.
   - **d.** shift toward the red end of the spectrum.

Answers: 1. c **2.** d **3.** d **4.** d

Chapter 20

MOTIONS OF THE EARTH, MOON, AND SUN

Chapter Checkin

- ❑ Knowing the Earth's and Moon's daily, monthly, and yearly positions
- ❑ Determining the phases of the Moon and understanding time measurements
- ❑ Applying Kepler's Laws of Planetary Motion

The motions of the Earth and heavenly bodies have been studied and documented for thousands of years. Time and position were told by the location of the Sun and the stars. Kepler's Laws of motion govern how these objects and the Earth move. The law of gravity applies to all bodies in the universe. As the amount of knowledge grew, so did the theories on planetary motions.

Motion of the Earth and Its Relation to Other Bodies

The motions of the Earth and other objects in the universe can be see as apparent of actual motions. These differences have affected how models of the solar system were viewed by scientists and other observers.

Apparent Motions in the Sky

As you observe celestial objects over the period of a few hours, they appear to move through the sky from east to west. This observation helped in the development of the geocentric model, in which all other objects revolved around the Earth. The Earth is really rotating around from west to east (or counter-clockwise, if viewed from above the North Pole).

Locating the Position of an Object in the Sky

Locating an object on the surface of the Earth requires two measurements, latitude and longitude. To find the position of an object in the sky, two measurements are also needed. These are the **altitude** (the angle above the horizon) and the **azimuth** (the direction to face). The point directly overhead is the **zenith** for the observer.

Geocentric and Heliocentric Models

The original model for the solar system was described by the Greek astronomer **Ptolemy** around the year A.D. 140. He said that the Earth was the center of the universe. This **geocentric** theory helped describe the daily motions of celestial objects around the Earth. It didn't explain the change of positions of planets in the sky over longer periods of time. He added the idea of **epicycles** to help explain this planetary motion. This idea of an Earth-centered universe lasted until the 1600s, when the Polish astronomer **Nicholas Copernicus** created the **heliocentric** model. This put the Sun at the center of the solar system. The Earth and other planets revolved around the Sun.

Actual Motions of the Earth

The Earth rotates on its axis, which is an imaginary line running through the poles. This line is not straight up and down, but has a 23.5° tilt. As the Earth revolves around the Sun, this axis always points in the same direction. This parallelism of the axis explains why Polaris is always above the North Pole, and why we have seasons.

One effect of the Earth's rotation is that we have day and night. The tilt of the Earth causes the concentration of the Sun's rays to move north and south over the course of a year. This creates seasons and the change in the amount of daylight and nighttime. If there were no tilt in the axis of rotation, we would have 12 hours each of day and night and no seasons. If the tilt were increased, we would have warmer summers and cooler winters.

The motions of celestial objects on a daily basis are a result of the Earth's rotation. The Earth orbits the Sun in a level plane in an elliptical path, with the Sun located at one of the focal points of the ellipse. The average distance between the Sun and the Earth is 150,000,000 km (93,000,000 miles). The closest point, or **perihelion,** occurs on January 1, when we are about 147,000,000 km (91,000,000 miles) away from the Sun. The **aphelion** position occurs on July 1, when the distance increases to about

153,000,000 km (95,000,000 miles) away. This 5 percent change in distance from the Sun has very little effect on the seasons.

Apparent Diameters of Celestial Objects

The actual diameter of a celestial object doesn't change significantly, but its **apparent diameter** can. This is the distance that we see it take up in the sky. As the object moves closer to us, it covers more of the field of view and appears larger. If the object moves away, it appears to get smaller. It really doesn't change its diameter, although it looks as if it does.

Evidence for Rotation

In 1851 the French physicist Jean Foucault proved that the Earth rotated through the use of a pendulum. If observed from space, pendulums stay in constant motion in one direction. When viewed on Earth, they appear to change their direction. Another piece of evidence for supporting the Earth's rotation theory is the Coriolis effect. This states that projectiles (wind, water currents, rockets, and so on) are deflected to the right in the Northern Hemisphere and to the left in the Southern Hemisphere.

Evidence for Revolution

By studying the positions of the stars in comparison to each other, it was observed that the stars appeared to change positions at different times of the year. This is because of **parallax.** The closer stars appear to move more due to the angle at which they are seen. This can be seen on a smaller scale as you drive along. Objects closer to you appear to move faster than objects in the distance. The parallax of stars wouldn't be a factor if the Earth didn't revolve around the Sun. The seasonal change in constellations also supports the idea that the Earth revolves.

Time Measurements

The Earth rotates once in about 24 hours. Because there are 360° of longitude around the Earth, the Earth rotates at an angular speed of 15°/hr. Linear speed (kilometers per hour) varies depending on your latitude. The equator has the largest circumference and the fastest speed, at about 1,670 km/hr. The speed at 42° latitude is about 1,300 km/hr; it is 0 km/hr at the North or South Pole. Solar time is told by the position of the Sun. As the Sun reaches its highest point for the day, it is at its **solar noon** position. This is also known as high noon or **local noon.**

The Earth is divided into 24 time zones, each an hour apart and 15° longitude wide. The time meridian goes through the middle of the time zone. There are some variances due to cities and islands falling in line with this meridian. If a common unit of time is needed, instead of local times, Greenwich Mean Time (GMT) is used. This is the time at the Prime Meridian in Greenwich, England. This prevents miscommunications due to time differences.

Daylight Saving Time was originally started to help save energy and to encourage people to spend time together outside. Clocks are set ahead one hour from spring to fall. The rest of the year is spent with the clocks back that hour, showing "standard time." You set your clocks according to the saying "Fall behind, spring ahead."

Another change that is needed is the day. The **International Date Line** was created for that purpose. It runs along the 180° longitude line, shifting occasionally for islands or countries. As you move westward across the line, a day is added. If you go eastward over the line, you go back a day.

Movements and Effects of the Moon

Orbiting the Earth is the Moon. As it moves around in its monthly path, several changes are affected by this motion. The phases of the Moon and tides in large bodies of water are tied to the position of the Moon.

Actual Motions of the Moon

The motion of the Moon around the Earth is not a flat or round path. The plane of the Moon's orbit is at about a 5° angle to the Earth's orbit. This helps to explain why the Moon is in different parts of the sky at different times and why eclipses are relatively rare. The path of the Moon's orbit is slightly eccentric. The position farthest away from the Earth is called the **apogee,** whereas the nearest point is called the **perigee.** The lunar month can be thought of in two ways. The period of revolution for the Moon is 27⅓ days. The time from one full Moon to the next is 29½ days. The reason for this difference is that the Earth is revolving around the Sun at the same time. The Moon takes about two days to catch up to the Earth to be in line for the full Moon phase. Figure 20-1 shows the positions of the Sun, Earth, and Moon during these times.

Figure 20-1 The monthly change in the positions of the Sun, Earth, and Moon.

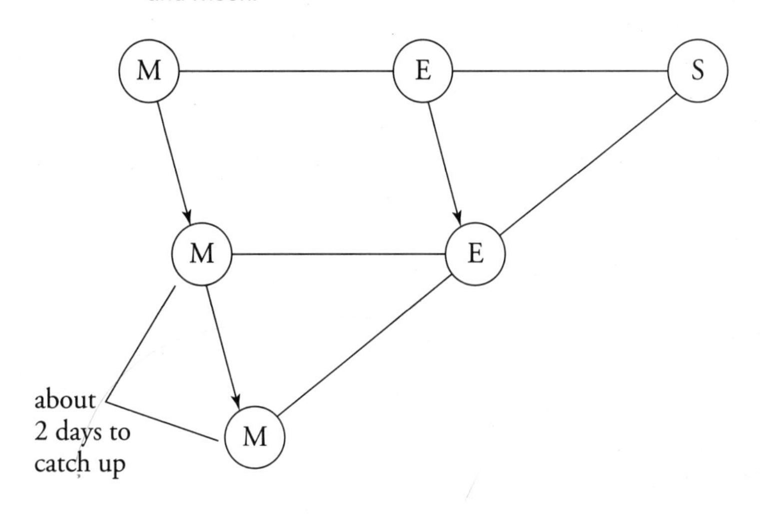

Phases of the Moon

As the Moon revolves around the Earth, you see it in different phases. Half of the Moon is always lit up. This is the side facing the Sun. When you look at the Moon, you see only half of it. You never see the side facing away from the Earth. The phase you see is a result of these two ideas.

As the Moon goes from the new Moon phase to the full Moon phase, more of it is getting brighter, or **waxing.** Think about when you *wax* something; it gets shiny and bright. You see the right side of the Moon first. Gradually, you see more of the Moon, moving toward the left side. About two weeks after the new Moon, you see a full Moon. Over the next two weeks, the Moon will diminish in size. The left side will continue to be seen, but the right side now is disappearing. The Moon is considered to be **waning** as it goes from the full Moon phase to the new Moon phase. It gets less full each night. This quote will help you to remember whether the Moon is waxing or waning by looking at the side most lit up: "Left lit leaving, right returning." Figure 20-2 shows the position of the Moon during each phase.

Figure 20-2 The Moon's position during each phase.

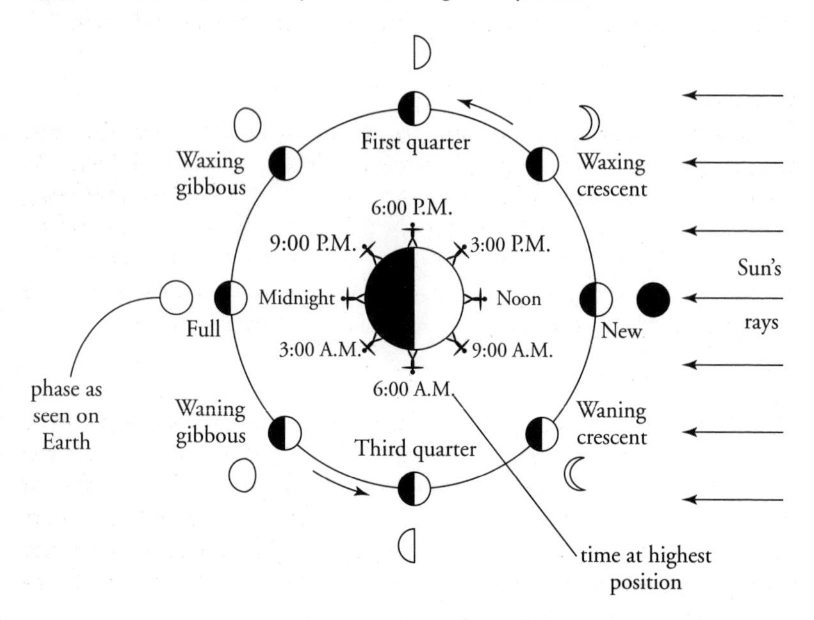

The first- and third-quarter positions are well named. You see only ¼ of the whole Moon during these times. The full Moon is not really a full Moon. You do see all of the lit up part, but you see only half of the Moon at a time. You can't actually see both sides of the moon.

Tides

The daily rising and falling of the ocean tides is caused by the Moon's gravitational force. The Earth and Moon pull on each other, but the liquid part of the Earth is most affected. The water layer is pulled to the side facing the Moon, forming a bulge. Another bulge forms on the opposite side of the Earth, as explained in Newton's Laws of Motion. These areas experience a high tide. The regions between the bulges are at low tide. The Earth rotates under the layer of water, causing an area to see the rise and fall of the tides. The Moon rises about 50 minutes later each day and the high tide for an area does the same.

Spring tides are times of extremely high and extremely low tides. This occurs when the Sun, Earth, and Moon are in a straight line. These are the times of the full Moon and new Moon phases. When the Sun, Earth, and Moon are at right angles to each other, as in the first-quarter and third-quarter phases of the Moon, **neap tides** are experienced. These have the smallest difference between the levels of the high-tide mark and the low-tide mark.

Eclipses

The shadows given off by the Earth and the Moon actually have two parts. You can see this by putting your hand near a desk in a brightly lit room. As you move your hand up and down, the image of the shadow goes from being well defined to having a fuzzy border. If you look closely at the fuzzy shadow, you can see a very dark shadow with a gray outline. The dark area is the **umbra** of the shadow and the gray area is the **penumbra** of the shadow. The shadow made is actually cone-shaped, due to the bending of light.

Lunar Eclipses

Although two eclipses can occur each month in theory, they don't occur that often. A **lunar eclipse** occurs during the full Moon phase, when the Moon passes in the Earth's shadow. If the Moon goes into the umbra of the Earth's shadow, a total eclipse occurs. A partial-shadow eclipse occurs when the Moon goes through the penumbra. When observing a lunar eclipse, the moon appears a coppery-red color due to the refraction of sunlight through Earth's atmosphere.

Solar Eclipses

A total eclipse of the Sun occurs when the Moon passes between the Earth and the Sun and the lunar shadow reaches the Earth. This can happen only during a new Moon phase, and usually during the local noontime for the small affected area. If the penumbra reaches Earth, a partial eclipse of the Sun occurs. If the umbra doesn't reach the Earth, an annular eclipse can be seen. Figure 20-3 shows the positions for both types of eclipses.

Figure 20-3 Eclipses and their positions.

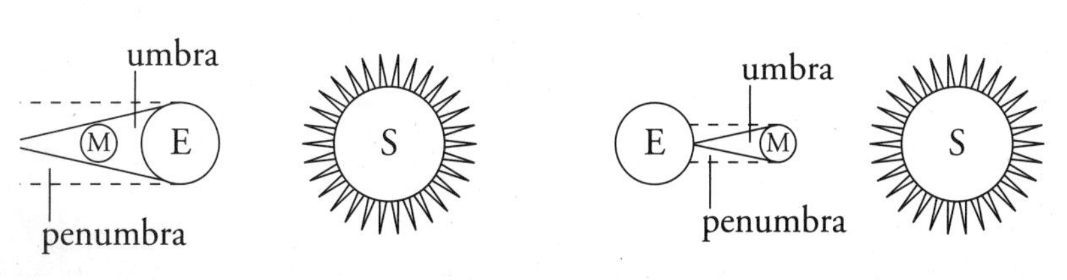

Kepler's Laws of Planetary Motion

The laws of motion that govern the planets were the result of the work of two men. Tycho Brahe, a Danish nobleman, lived on an island in the 1500s and built an observatory. He kept very careful records and measurements of the positions of stars and planets over a 20-year period. He was a brilliant observer, but didn't get a chance to do much with his data. A German mathematician, **Johannes Kepler,** was Brahe's apprentice. Kepler inherited all of the data (there are other versions to this story) and developed the three **laws of planetary motion** that we still use today, which he could only have done with that many years of planetary observations to support his theories. The **First Law of Planetary Motion** states that planets move in elliptical orbits, with the Sun at one of the focal points. The other focal point is empty space. The closest point to the Sun is the **perihelion** point. The position farthest away from the Sun is the **aphelion** ("a" = away) point. The **Second Law of Planetary Motion** says that planets sweep out equal areas in equal time. As the planet gets further away from the Sun, the force of gravity is weaker and the planet slows down. The speed of the planet increases as it approaches the Sun. These two laws are illustrated in Figure 20-4.

Figure 20-4 Planetary laws.

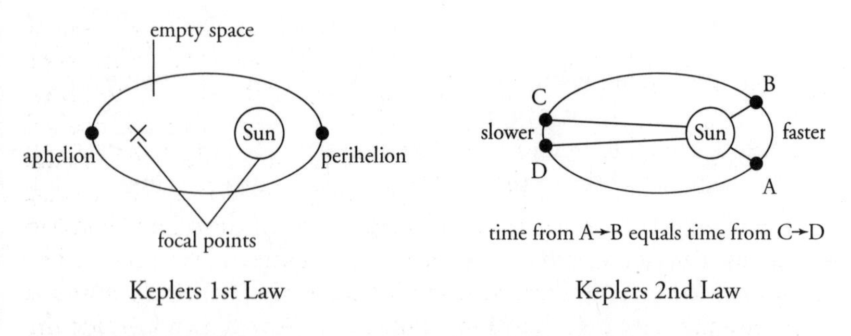

The **Third Law of Planetary Motion** is also called the **harmonic law,** which involves the period of revolution (P) and the distance from the Sun (D). The formula for this law is $P^2 = D^3$, where the period is measured in years and the distance is measured in astronomical units. Basically, it says that the farther you are from the Sun, the longer it takes for you to revolve around it.

Galileo, who lived at about the same time as Kepler and Copernicus, made observations with his telescope of the Moon, Venus, Jupiter, and Saturn. His observations helped to support their theories. This didn't sit well with the Catholic Church, which supported the geocentric model. Galileo was put under house arrest, where he eventually died in 1642. It wasn't until about 350 years later that he was officially apologized to by the church.

Gravity

Sir Isaac Newton's **universal law of gravity** helped to explain how planets were kept in motion around the Sun as described by Kepler's laws. **Newton** showed that the force of gravity between two objects depends directly on the masses of the two objects and is inversely related to the square of the distance between the objects. In other words, as the masses increase, the force of gravity increases. As the distance between the objects increases, the force of gravity decreases. Newton's law of universal gravitation is described by the following formula:

$$F = \frac{Gm_1 m_2}{d^2}$$

Space Travel

The path that led man to the Moon had several stages. The *Luna* series of space probes were sent up by the Soviet Union. The goal of these missions was to orbit the Moon. The *Pioneer* series did the same for the United States. The next set of probes from the *Ranger* missions took pictures and crashed on the Moon. The *Surveyor* probes did the same, but landed safely on the Moon. The *Mercury* missions carried astronauts around the Earth and returned them. The *Gemini* spacecraft carried two astronauts and had two objectives. The first was to test humans in a weightless environment. The second was to meet and dock with another spaceship. The *Apollo* missions were the culmination of these programs. They allowed men to reach the Moon and return safely to Earth. Rock and soil samples were returned to Earth (about 380 kg) and instruments were left behind. These measured Moon quakes, magnetic fields, solar winds, and gases on the Moon's surface. More recent space programs include the Space Shuttle launches, unmanned missions to Mars and the creation of the International Space Station.

Chapter Checkout

Q&A

1. In our solar system, the orbits of the planets are best described as
   a. circular, with the planet at the center.
   b. circular, with the Sun at the center.
   c. elliptical, with the planet at one of the foci.
   d. elliptical, with the Sun at one of the foci.

2. Based on observations made in the Northern Hemisphere, which statement is the best supporting evidence that the Earth rotates on its axis?
   a. The stars appear to follow daily circular paths around Polaris.
   b. The apparent solar diameter varies throughout the year.
   c. The length of the daylight period varies throughout the year.
   d. The seasons (spring, summer, fall, and winter) repeat in a cyclic pattern.

3. Some stars that can be seen in New York State on a summer night cannot be seen on a winter night. This fact is a result of the
   a. rotation of the Earth on its axis.
   b. rotation of the stars around Polaris.
   c. revolution of Polaris around the Earth.
   d. revolution of the Earth around the Sun.

4. If the Earth's rate of rotation decreased, there would be an increase in the
   a. length of the seasons.
   b. Sun's angle of insolation at noon.
   c. number of observable stars at night.
   d. length of time for one Earth day.

Answers: 1. d **2.** a **3.** d **4.** d

Chapter 21
DATING OF ROCKS

Chapter Checkin

- ❑ Understanding the order of events in a series of rock layers
- ❑ Knowing about index fossils
- ❑ Determining the absolute age of rocks, fossils, and events

Rocks, fossils, and events can be dated in two ways. They can be compared to each other in a relative manner and listed in a sequence. The actual date of an event can also be determined. These events are recorded in the rock record and show the existence of life and events from the past.

Uniform Processes

One of the assumptions made in determining the age of rocks and fossils is that processes that occurred in the past are the same as what happens today. The forces that erode rocks and the manner in which rocks and mountains are formed haven't changed over time. These events will also take place in the same way in the future. This is part of James Hutton's concept of uniformitarianism, which he proposed in 1795.

Relative Dating of Rocks and Events

The age of rocks, fossils, and events in comparison to each other is called relative dating. The sequence of events is investigated. There are several laws that help to determine this order. The law of superposition states that the youngest rock layers are on top. The idea behind this is similar to clothes in a hamper. The most recent additions are on the top of the pile. Superposition is related to the idea of original horizontality. This states that the sedimentary rock layers are made in horizontal, flat layers, so if they are found in anything other than this manner a more recent event has occurred. If an igneous intrusion cuts across the existing rock layers, it is younger than the layers that it cuts across. This is the law of cross-cutting. It makes sense that the rock layers have to be in place before the intrusion can cut across them. In conglomerates, which are sedimentary rocks, pieces of other existing rocks are put together to form the new rock. This is the main idea in the law of included fragments. The pieces of rock are older than the rock they are found in. An **unconformity** is a break or gap in the rock record. These unconformities occur when rock layers under water are uplifted and emerge out of the water. The rock is then exposed to erosional forces, removing the top layers. The rock layers remaining are then submerged again under water, where newer layers of rock are formed on top. The region then again gets raised above sea level, where we can study the layers. A sample cross-section can been seen in Figure 21-1.

Figure 21-1 Cross-section rock layers in the Earth's crust. Line XY is a fault.

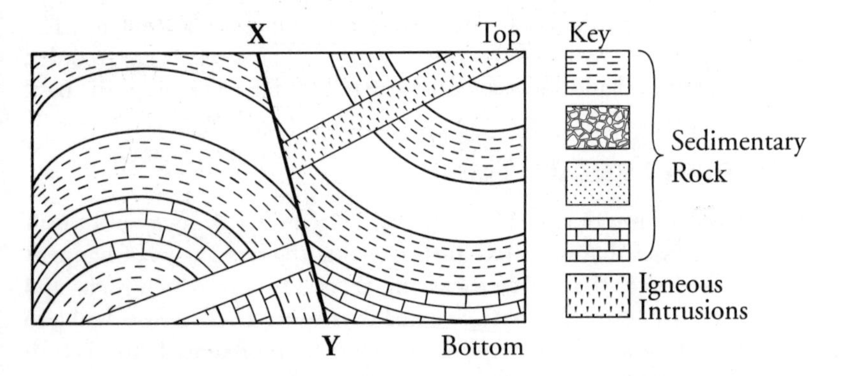

The order of events for Figure 21-1 is formation of sedimentary layers, folding of layers, igneous intrusion, faulting along X-Y, and erosion at the surface.

Correlation of Rock Layers and Events

Correlation is the matching of rock layers to each other between two different areas. The layers are looked at in relative terms compared to each other. The easiest way to match the layers is to "walk the outcrop." This means that you look at cross-sections of the layers to be compared. Similar rock patterns and index fossils help to connect layers. Deposits of volcanic ash is another good way to connect layers together. Unconformities can make it more difficult to connect layers to each other because there is missing information. Figure 21-2 shows two rock outcrops. Rock layers B and X can be connected together.

Figure 21-2 Rock outcrops.

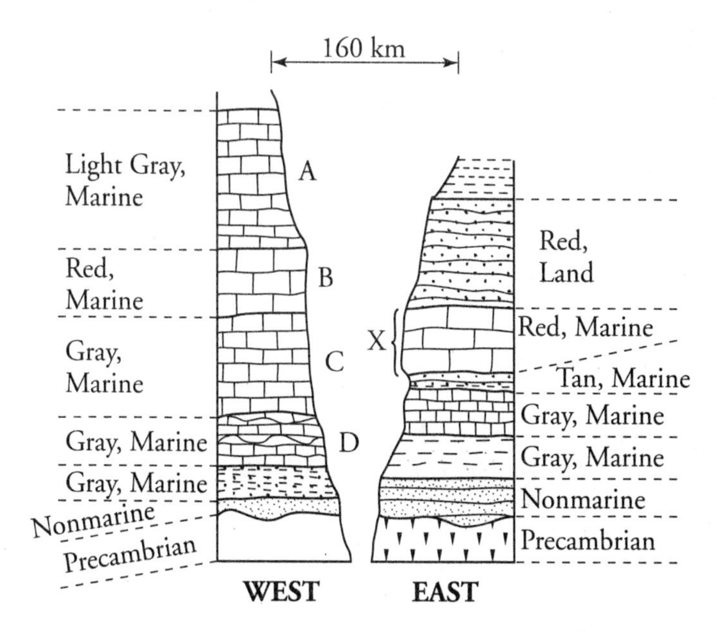

Index Fossils

Index fossils help to correlate rock layers to the same time period. They are fossils of a creature that lived for a short period of time over a widespread area. Trilobites are an example of an organism that is used as an index fossil. **Key beds** are also used to connect rock layers. They are single rock layers that have the same characteristics as an index fossil. Volcanic eruptions fall into this category. The eruptions don't last very long and ash from the eruption is spread over a wide area.

Absolute Dating and Radioactive Decay

The age of an event or an organism can be determined in terms of years. The method for dating these occurrences varies depending on how long ago it was.

Tree Rings

On a shorter scale, trees can be used for determining temperature and rainfall. This comes from studying the tree rings. The alternating light and dark rings are made over the period of a year. Scientists have been able to study wood from Native American ruins in the Southwestern United States to determine the climate from about 4,000 years ago. The oldest living trees are the bristlecone pine trees in the Sierra Nevada Mountains in California. One tree is more than 4,600 years old.

Varves

Varves can be studied as well. These formations are made from sediments that show yearly cycles. They can be found in any lake or ocean, especially in glacial lakes. During the summer, meltwater from the glaciers carries sand particles into the lake, which settle to the bottom. This creates a thick, light-colored, sandy layer. Summer breezes keep the water mixed, so smaller particles don't settle out. During the winter, the lake freezes, preventing the winds from mixing the water column. There is also little runoff from the glaciers. The silt is now unaffected by these factors, which allows it to sink to the bottom. The result is a thinner, dark-colored layer of clay particles. Varves can show climate conditions back to about 15,000 years.

Radioactive Elements

All living creatures contain some **radioactive elements.** A radioactive atom gives off alpha particles (similar to Helium atoms—two protons, two neutrons), beta particles (high-energy electrons), and gamma rays (like X-rays) as it decays. When alpha particles are given off, the element changes to a new element and becomes lighter. This decay continues until it becomes a stable atom or a nonradioactive atom. For example, uranium-238 decays into lead-206.

Half-Life

The **half-life** of a radioactive element is the time it takes for half of the radioactive atoms to decay into the stable end product. This rate can vary greatly for different substances. Figure 21-3 shows some of the more commonly used **isotopes.**

Figure 21-3 Radioactive decay data.

RADIOACTIVE ISOTOPE	DISINTEGRATION	HALF-LIFE (years)
Carbon-14	$C^{14} \longrightarrow N^{14}$	5.7×10^3
Potassium-40	$K^{40} \longrightarrow Ar^{40}$, $K^{40} \longrightarrow Ca^{40}$	1.3×10^9
Uranium-238	$U^{238} \longrightarrow Pb^{206}$	4.5×10^9
Rubidium-87	$Rb^{87} \longrightarrow Sr^{87}$	4.9×10^{10}

The half-life of a substance can be graphed in two ways. One graph shows the amount of original radioactive remaining versus half-life. The other graph plots the amount of end product against time. Both of these ways can be seen in Figure 21-4.

Figure 21-4 A substance's half-life.

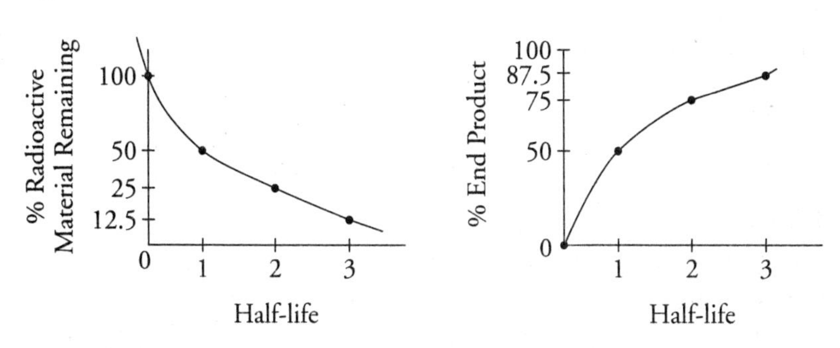

Radioactive Dating

The age of a fossil or a rock can be determined by looking at the radioactive isotopes. In animals and plants, there is a consistent amount of C^{14}, which is an isotope of C^{12}, until the organism dies. Then the decay process begins. Carbon is the element that makes up the structure of most of our molecules. This type of dating has been used since the late 1940s. The half-life of C^{14}

is 5,700 years. This method can be used to date objects back about 100,000 years. Other methods use radioactive potassium, uranium, and rubidium. These have much longer half-lives and are used for dating very old rocks. Because the half-life is so long for uranium and rubidium, these isotopes cannot be used for dating relatively young rocks. Potassium has a shorter half-life and is very common in all rocks in the Earth's crust (such as mica and feldspar). It can be used to date rocks as young as 50,000 years.

Chapter Checkout

Q&A

1. Unless a series of sedimentary rock layers has been overturned, the bottom layer usually

- **a.** contains fossils.
- **b.** is the oldest.
- **c.** contains the greatest variety of minerals.
- **d.** has the finest texture.

2. Unconformities (buried erosional surfaces) are good evidence that

- **a.** many life forms have become extinct.
- **b.** the earliest life forms lived in the sea.
- **c.** part of the geologic rock record is missing.
- **d.** metamorphic rocks have formed from sedimentary rocks.

3. Radioactive carbon-14 dating has determined that a fossil is 5.7×10^3 years old. What is the total amount of the original C^{14} still present in the fossil?

- **a.** 0%
- **b.** 25%
- **c.** 50%
- **d.** 75%

4. In which way are index fossils and volcanic ash deposits similar?

- **a.** Both normally occur in nonsedimentary rocks.
- **b.** Both can usually be dated with radiocarbon.
- **c.** Both often serve as geologic time markers.
- **d.** Both strongly resist chemical weathering.

Answers: 1. b **2.** c **3.** c **4.** c

Chapter 22

EVOLUTION OF LIFE ON EARTH

Chapter Checkin

- ❑ Knowing the geologic time scale and important events in each period
- ❑ Understanding the development of the variety of life forms in the Earth's history
- ❑ Determining how fossils form

Most of the time that the Earth has been around it was without life. When life appeared, it started in the oceans and eventually evolved onto land. The explosion of numbers and varieties of life forms has come over a short, recent time period, geologically speaking.

Geologic Time Scale

The geologic time table is a summary of major events that have occurred during the history of the Earth. These are preserved in the rock record and are divided into eons, eras, periods, and epochs. The beginning or end of each of these time periods is marked by an event, not by a specific amount of time. These events could be the disappearance or appearance of an organism or a geologic event. As creatures appear or disappear from the Earth, other creatures are affected. Some flourish, some perish, and others evolve in response to their environment.

Precambrian Eon: 4.6 bya–570 mya

During most of this time, the Earth was cooling and the atmosphere was developing. Life was sparse, with only about 40 known fossil locations. Most of the fossils found from this time period were stromatolites. Important mineral deposits were formed during this time period. These include iron deposits around Lake Superior; magnetite and ilmenite ores in the

Adirondack Mountains; nickel deposits in Ontario, Canada; uranium ores at Great Bear Lake in Northwest Canada; and gold ores in South Africa. At least four mountain-building episodes occurred during this time. These events are called **orogenies.** Rocks from this time period can be found in the Adirondack Mountains, under New York City, in the Piedmont region (between the Appalachian Mountains and the Atlantic Ocean), the core area of the Rocky Mountains, and at the bottom of the Grand Canyon in Arizona.

Archeozoic Era: 4.6 bya–2.5 bya

Fossils from this time period are rare because most of the rocks that formed during the Archeozoic Era were igneous and metamorphic in origin. Most of the creatures that lived during this time didn't have skeletons or hard shells. This makes it difficult for the organism to be preserved in the rock record.

Proterozoic Era: 2.5 bya–570 mya

The emergence of simple plants and worms in the oceans occurred during this era. There was no life on land at this time.

Phanerozoic Eon: 570 mya–present

During this time period, the amount and variety of life forms exploded. Many life forms also disappeared during several mass extinctions.

Paleozoic Era: 570 mya–230 mya

Land and ocean plants appeared during the Paleozoic Era. Simple animals were also developing and evolving. This era started with an abundance of sea life and ended with the extinction of many marine animals. The Paleozoic Era is subdivided into the following periods:

- **Cambrian:** 544 mya–490 mya, animals with hard parts; lots of trilobites and brachiopods.
- **Ordovician:** 490 mya–443 mya, increase in the numbers of species; graptolites are important index fossils for this period; early vertebrates appeared.
- **Silurian:** 443 mya–418 mya, eurypterids thrived in the ocean; plants and animals such as spiders, millipedes, insects, and scorpions started to live on land.
- **Devonian:** 418 mya–362 mya, *the Age of Fishes;* coral reefs reach their peak; fish (jawless, covered in plates) started to appear; lungfish (which

gave rise to amphibians) appeared; land plants started to increase in variety (spores, seed-producing, trees with bark, and conifers).

- **Carboniferous (Mississippian, Pennsylvanian):** 362 mya–290 mya, the development of crinoids and foraminfera were the high points of Mississippian period; the coal beds of Pennsylvania, Ohio, West Virginia, Indiana, and Illinois were created during the Pennsylvanian Period; reptiles also appeared during this time; insects thrived, especially cockroaches—the Pennsylvanian period is called the *Age of Cockroaches.*
- **Permian:** 290 mya–251 mya, mostly a dry climate, except for a major ice age in the Southern Hemisphere.

Mesozoic Era: 230 mya–65 mya

This is the time of dinosaurs. They thrived during this time and ruled the Earth. They disappeared, for the most part, at the end of this era. More than 50 percent of the plant and animal species were wiped out during the extinction 65 mya. The climate during this time was mild and the poles were free of ice.

- **Triassic:** 251 mya–206 mya, small dinosaurs evolved; in the ocean, ammonites appeared (index fossil).
- **Jurassic:** 206 mya–142 mya, birds and mammals appeared.
- **Cretaceous:** 142 mya–65 mya, large dinosaurs ruled the land; flowering plants and deciduous trees appeared.

Cenozoic Era: 65 mya–present, the Age of Mammals

Several ice ages, separated by warm periods, were recorded during this time. Humans and modern plants evolved during this time.

- **Tertiary:** 65 mya–1.6 mya, grasses appeared; grazing and carnivorous mammals flourished; birds evolved.
- **Quarternary:** 1.6 mya–present, humans arrived on Earth; some species became extinct from human interference (hunting, elimination of breeding grounds), and many more are in danger of becoming extinct. More than 88 percent of the deforestation in the Amazon rain forests has occurred since 1980. Considering the fact that about ¼ of all pharmaceuticals are derived from rain-forest plants (some of which we have not even classified yet!), man has had a major impact on other life forms on Earth.

Evolution of the Earth and Its Atmosphere

The Earth began to form about 4.6 billion years ago. It was heated by radioactive elements decaying and releasing energy and from friction of the movement of materials. These materials separated themselves by density, creating the layers that exist today. The inner layers are composed of iron and nickel, whereas the outer layers are less dense silicates. About 4.2 bya, the crust formed and plate tectonic activity began. Gases that escaped through the crust during volcanic eruptions is called **outgassing** and helped to form the early atmosphere. As the Earth cooled, precipitation fell and oceans formed. This allowed sedimentary rocks to begin to form.

Development of the Atmosphere

As the Earth cooled and solidified about 4 billion years ago, the beginnings of life began. About 3.5 bya, carbon fixing began, which is the basis for all life forms. Carbon fixing is the conversion of carbon from CO_2 to a carbohydrate form. Organisms using CO_2 and chemosynthesis of sulfur started to appear. The transition of the atmosphere to having O_2 covered the time period from 3.4 bya to –1.5 bya. Once free oxygen was in the air and water, other life forms appeared.

Major Geologic Events

Several major geologic episodes have created the landscape that we see today on the Earth's surface. Mountain-building events are constantly being counteracted by erosional forces. The motions of the plates have had a deep impact on the landscapes by creating mountains and altering the climate. As the plates move to different areas on the Earth, the climate changes, affecting the plant and animal life and the weather patterns that existed.

- **Ordovician Period:** the Taconic Orogeny (a piece of Africa broke off and collided with North America) created the Green Mountains in Vermont and the Taconic Mountains in New York.

- **Devonian Period:** the Acadian Orogeny (North America colliding with Africa) created the Appalachian Mountains and the White Mountains in New Hampshire.

- **Mississippian and Pennsylvanian periods:** the Allegheny Orogeny (uplift) created the middle and southern Appalachian Mountains and the beginnings of the Rocky Mountains; Pangaea forms.
- **Triassic Period:** Pangaea begins to break up into Laurasia and Gondwanaland.
- **Cretaceous Period:** Laurasia and Gondwanaland break up into the continents we see today.
- **Cenozoic Era:** the Colorado Plateau is uplifted and the Colorado River starts creating the Grand Canyon; faulting creates the Nevada Range and Sierra Nevadas in California; lava flows create the Columbia Plateau in Washington, Oregon, Idaho, and California and the Cascade Mountains in the west; other mountain ranges built up during this time were the Himalayas in Asia, the Alps in Europe, and the Andes in South America; ice sheets covered 25 percent of the Earth during the Pleistocene Epoch, as opposed to about 10 percent now.

All of these events can be seen in Figure 22-1.

Evolution of Life on Earth

The rock records show the appearance and disappearance of organisms. Some organisms are replaced by new, different organisms. In other cases, the organisms show changing or evolving patterns.

Theory of Evolution

The theory of evolution is the process of change that produces new life forms over geologic time. In 1859 Charles Darwin, a British naturalist, put forth his theory of **natural selection.** He studied many creatures while traveling the world aboard the *H.M.S. Beagle,* especially finches off of the coast of South America. His premise was that organisms that survive pass along their traits to their offspring. These successors are then best adapted for surviving in their environment. Recent studies and theories have found that some organisms just "appear" in the fossil record. This is known as **punctuated equilibrium.**

Figure 22-1 Geologic history and events.

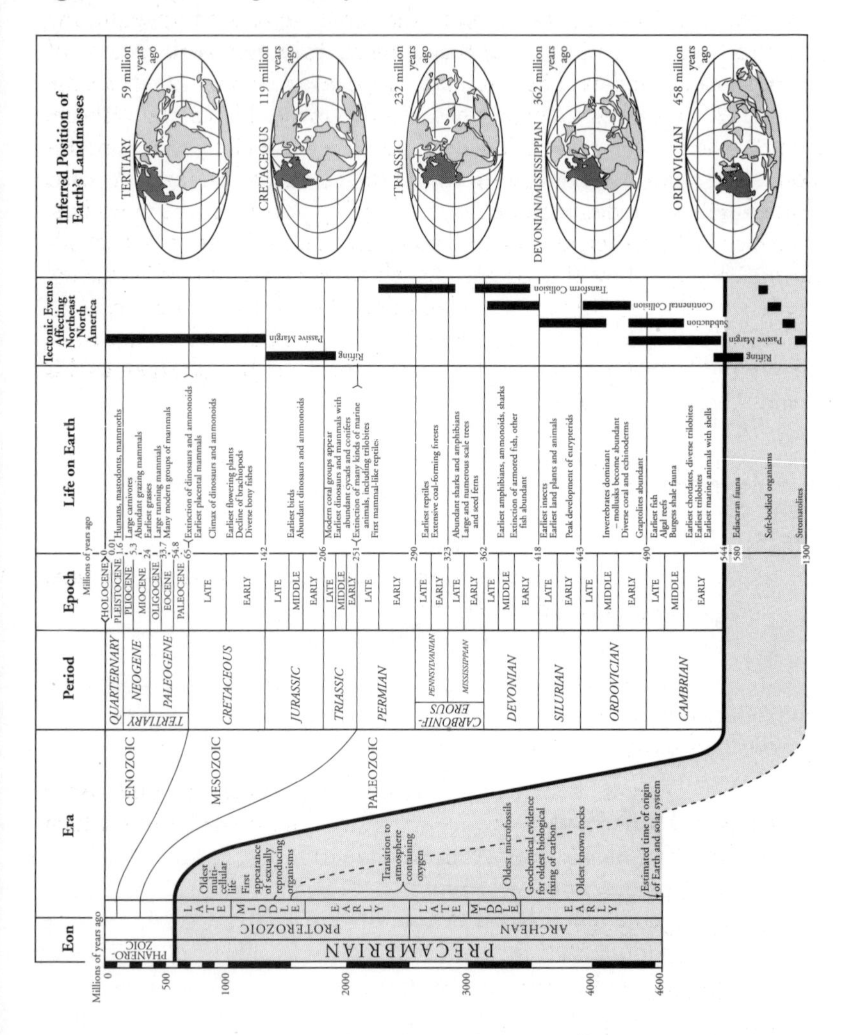

Fossils

Fossils are evidence of earlier life that is preserved in a rock. These can be shells, bones, petrified trees, impressions made by plant leaves, footprints, or burrows made by worms. Original remains can be preserved in ice, as in permafrost conditions. Amber is the sticky sap, containing resin, from pine trees that hardens. Insects and other small items are trapped in the sap and become fossils. Some fossils result from the replacement of the organism's remains. The soft parts disappear and minerals replace the hard parts. This is also how petrified wood is made. Molds and casts are formed when a shell or bone dissolves out of a rock. The hollow impression is the mold. New minerals fill the mold, which becomes the cast. **Trace fossils** are evidence of life that are fossilized. These can be footprints, burrows, trails, tracks, and so on. Fossils are good indicators of past climates. They can show that organisms, which have certain environmental needs, lived during a certain period of time. Microfossils are important in the area of oil exploration. Deep core samples are taken and compared. If oil is found in one area, rock layers in other areas can be correlated to this area by studying the microfossils.

Chapter Checkout

Q&A

1. The Geologic Time Scale has been subdivided into a number of time units called periods on the basis of

a. fossil evidence.
b. rock thickness.
c. rock types.
d. radioactive dating.

2. A layer of shale is located below Devonian limestone and above Cambrian sandstone. Overturning has not occurred. During which period was the shale most likely deposited?

a. Silurian
b. Carboniferous
c. Triassic
d. Tertiary

3. When did the dinosaurs become extinct?

a. Before the earliest birds
b. Before the earliest mammals
c. At the end of the Cretaceous Period
d. At the end of the Cambrian Period

4. Trilobite fossils from different time periods show small changes in appearance. These observations suggest that the changes may be the result of
   a. evolutionary development.
   b. a variety of geologic processes.
   c. periods of destruction of the geologic record.
   d. the gradual disintegration of radioactive substances.

Answers: 1. a **2.** a **3.** c **4.** a

CQR REVIEW

Use this CQR Review to practice what you have learned in this book. After you work through the review questions, you're well on your way to understanding Earth Science.

Chapter 1

1. Which procedure is an example of classifying observed data?

a. grouping stars by brightness
b. graphing temperature versus time for a particular date
c. photographing the cloud cover for a location for 1 week
d. measuring the angle of Polaris from two different locations

2. Which event is most predictable?

a. The Sun rises.
b. An earthquake occurs.
c. A meteorite falls to Earth.
d. Coral fossils are found on mountaintops.

3. Ocean tides are best described as

a. unpredictable and cyclic.
b. unpredictable and noncyclic.
c. predictable and cyclic.
d. predictable and noncyclic.

4. A list of three observed relationships is shown here.

Erosional rate = depositional rate

Amount of insolation = amount of terrestrial radiaiton

Rate of condensation = rate of evaporation

In which situation would each relationship exist?

a. when a cyclic change occurs
b. when a change of state occurs
c. when dynamic equilibrium is reached
d. when global warming ceases and global cooling begins

5. A pebble has a mass of 35 grams and a volume of 14 cubic centimeters. What is its density?

a. 0.4 g/cm^3
b. 2.5 g/cm^3
c. 490 g/cm^3
d. 4.0 g/cm^3

Chapter 2

6. Earth's hydrosphere is best described as the

a. solid layer of Earth.
b. liquid layer of Earth.
c. magma layer located below the Earth's stiffer mantle.
d. gaseous layer extending several hundred kilometers from Earth into space.

7. Measurements taken from space show the Earth to be

a. greatest in diameter at the Equator.
b. greatest in diameter at the Poles.
c. a perfect sphere.
d. pear shaped.

8. To an observer on the Moon, the Earth in full phase would appear to be shaped like

a. an egg.
b. a basketball.
c. a pear.
d. a football.

9. In which group are the spheres of the Earth listed in order of increasing density?

a. atmosphere, hydrosphere, lithosphere
b. hydrosphere, lithosphere, atmosphere
c. lithosphere, hydrosphere, atmosphere
d. lithosphere, atmosphere, hydrosphere

10. In which atmospheric layer is most water vapor found?

a. troposphere
b. stratosphere
c. mesosphere
d. thermosphere

Chapter 3

11. As a ship crosses the Prime Meridian, the altitude of Polaris measured from the ship is 50°. What is the ship's location?

a. 0° latitude 50° east longitude
b. 0° latitude 50° west longitude
c. 50° north latitude 0° longitude
d. 50° south latitude 0° longitude

12. Which quantity has both magnitude and direction?

a. relative humidity of 96%
b. westerly wind of 10 knots
c. pressure of 1012 mb
d. mass of 20 g

13. At which location will the highest altitude of the star Polaris be observed?

a. Equator
b. Tropic of Cancer
c. Arctic Circle
d. central New York State

14. Which statement is true about an isoline on an air temperature field map?

a. It represents an interface between high and low barometric pressures.
b. It indicates the direction of maximum insolation.
c. It increases in magnitude as it bends southward.
d. It connects points of equal air temperature.

15. When the time of day for a certain ship at sea is 12 noon, the time of day at the Prime Meridian (0° longitude) is 5 P.M. What is the ship's longitude?

a. 45° E
b. 45° W
c. 75° E
d. 75° W

Chapter 4

16. Which mineral leaves a green-black powder when rubbed against an unglazed porcelain plate?

a. galena
b. graphite
c. hematite
d. pyrite

17. Minerals are found in many different rocks. Which two rocks are primarily composed of a mineral that bubbles with acid?

a. limestone and marble
b. granite and dolostone
c. sandstone and quartzite
d. slate and conglomerate

18. Which mineral scratches dolomite and is scratched by olivine?

a. galena
b. quartz
c. potassium feldspar
d. muscovite mica

19. An unidentified mineral that is softer than calcite exhibits a metallic luster and cubic cleavage. This mineral most likely is

a. galena.
b. pyrite.
c. halite.
d. pyroxene.

20. Which statement can be made about the minerals plagioclase feldspar, gypsum, biotite mica, and talc?

a. These minerals have the same chemical and physical properties.
b. These minerals have different chemical properties, but similar physical properties.
c. These minerals have different physical and chemical properties, but identical uses.
d. The physical and chemical properties of these minerals determine how humans use them.

Chapter 5

21. Which sedimentary rock is most likely to be changed to slate during regional metamorphism?

a. breccia
b. conglomerate
c. dolostone
d. shale

22. Which process most likely formed a layer of the sedimentary rock, gypsum?

a. precipitation from seawater
b. solidification of magma
c. folding of clay-sized particles
d. melting of sand-sized particles

23. Which igneous rock, when weathered, could produce sediment composed of the minerals potassium feldspar, quartz, and amphibole?

a. gabbro
b. granite
c. andesite
d. basalt

24. Which physical characteristic best describes the rock phyllite?

a. glassy texture with gas pockets
b. clastic texture with angular fragments
c. bioclastic texture with cemented shell fragments
d. foliated texture with microscopic mica crystals

25. When granite melts and solidifies, it becomes

a. a sedimentary rock.
b. an igneous rock.
c. a metamorphic rock.
d. sediments.

Chapter 6

26. Landscapes will undergo the most chemical weathering if the climate is

a. cool and dry.
b. cool and wet.
c. warm and dry.
d. warm and wet.

27. Which agent of erosion is responsible for cutting most V-shaped valleys into bedrock?

a. surface winds
b. running water
c. glacial ice
d. ocean waves

28. Glaciers often form parallel scratches and grooves in bedrock because glaciers

a. deposit sediment in unsorted piles.
b. deposit rounded sand in V-shaped valleys.
c. continually melt and refreeze.
d. drag loose rocks over Earth's surface.

29. A rock will weather faster after it has been crushed because its

a. volume has been increased.
b. surface area has been increased.
c. density has been decreased.
d. molecular structure has been altered.

30. The occurrence of a U-shaped valley indicates that the area has most likely been eroded by

a. a glacier.
b. a stream.
c. waves.
d. wind.

Chapter 7

31. Where is the most deposition likely to occur?

a. on the side of a sand dune facing the wind
b. at the mouth of a river, where it enters an ocean
c. at the site where glacial ice scrapes bedrock
d. at the top of a steep slope in a streambed

32. When the velocity of a stream suddenly decreases, the sediment being transported undergoes an increase in

a. particle density.
b. erosion.
c. deposition.
d. mass movement.

33. Outwash plains are formed as a result of deposition by

a. landslides.
b. ocean waves.
c. winds from hurricanes.
d. meltwater from glaciers.

34. Which statement best describes sediments deposited by glaciers and rivers?

a. Glacial deposits and river deposits are both sorted.
b. Glacial deposits are sorted, and river deposits are unsorted.
c. Glacial deposits are unsorted and river deposits are sorted.
d. Glacial deposits and river deposits are both unsorted.

35. A stream with a velocity of 100 cm/sec flows into a lake. Which sediment-size particles would the stream most likely deposit first as it enters the lake?

a. boulders
b. cobbles
c. pebbles
d. sand

Chapter 8

36. Landscapes with horizontal bedrock structure, steep slopes, and high elevations are classified as

a. plateau regions.
b. plain regions.
c. lowland regions.
d. mountain regions.

37. In which type of landscape are meandering streams most likely found?

a. regions of waterfalls
b. gently sloping plains
c. steeply sloping hills
d. V-shaped valleys

38. The boundaries between landscape regions are usually determined by the location of

a. state boundaries.
b. major cities.
c. population density.
d. well-defined surface features.

39. Which evidence could be used to help classify a landscape region as a plateau?

a. rounded peaks
b. trellis drainage pattern
c. V-shaped river valleys
d. horizontal rock structure

40. The major landscape regions of the United States are identified chiefly on the basis of

a. similar surface characteristics.
b. similar climatic conditions.
c. nearness to major mountain ranges.
d. nearness to continental boundaries.

Chapter 9

41. Which set of conditions would produce the most runoff of precipitation?

a. gentle slope and permeable surface
b. gentle slope and impermeable surface
c. steep slope and permeable surface
d. steep slope and impermeable surface

42. Soil composed of which particle size usually has the greatest capillarity?

a. silt
b. fine sand
c. coarse sand
d. pebbles

43. Soil with the greatest porosity has particles that are

a. poorly sorted and densely packed.
b. poorly sorted and loosely packed.
c. well sorted and densely packed.
d. well sorted and loosely packed.

44. Which surface soil conditions allow for the most infiltration of rainwater?

a. steep slope and permeable soil
b. steep slope and impermeable soil
c. gentle slope and permeable soil
d. gentle slope and impermeable soil

45. Which water budget condition exists when precipitation is less than potential evapotranspiration and storage is depleted?

a. moisture surplus
b. moisture recharge
c. moisture usage
d. moisture deficit

Chapter 10

46. The Earth's outer core and inner core are both inferred to be

a. liquid.
b. solid.
c. composed of a high percentage of iron.
d. under the same pressure.

47. What is the inferred temperature at the boundary between Earth's stiffer mantle and outer core?

a. 2,500 °C
b. 4,500 °C
c. 5,000 °C
d. 6,200 °C

48. According to the plate tectonics theory, the Peru-Chile Trench and the Andes Mountains formed along the west coast of South America because the South American Plate

a. collided with the Nazca Plate.
b. collided with the North American Plate.
c. slid away from the Nazca Plate.
d. slid away from the North American Plate.

49. The theory of continental drift suggests that the

a. continents moved due to changes in the Earth's orbital velocity.
b. continents moved due to the Coriolis effect caused by the Earth's rotation.
c. present day continents of South America and Africa are moving toward each other.
d. present day continents of South America and Africa once fit together like puzzle parts.

50. Igneous rocks on the ocean floor that have an alternating pattern of magnetic orientation provide evidence that

   a. mountains are rising.
   b. the seafloor is spreading.
   c. the Earth was struck by meteorites.
   d. ocean tides are cyclic.

Chapter 11

51. The study of how seismic waves change as they travel through Earth has revealed that

   a. P-waves travel more slowly than S-waves through Earth's crust.
   b. Seismic waves travel more slowly through the mantle because it is very dense.
   c. Earth's outer core is solid because P-waves are not transmitted through this layer.
   d. Earth's outer core is liquid because S-waves are not transmitted through this layer.

52. An earthquake's P-wave arrived at a seismograph station at 02 hours 40 minutes 0 seconds. The earthquake's S-wave arrived at the same station 2 minutes later. What is the approximate distance from the seismograph station to the epicenter of the earthquake?

   a. 1,100 km
   b. 2,400 km
   c. 3,100 km
   d. 4,000 km

53. How long would it take for the first S-wave to arrive at a seismic station 4,000 km away from the epicenter of an earthquake?

   a. 5 min 40 sec
   b. 7 min 0 sec
   c. 12 min 40 sec
   d. 13 min 20 sec

54. A huge undersea earthquake off of the Alaskan coastline could produce a

   a. tsumani.
   b. cyclone.
   c. hurricane.
   d. thunderstorm.

55. An earthquake's P-wave traveled 4,800 kilometers and arrived at a seismic station at 5:10 P.M. At approximately what time did the earthquake occur?

a. 5:02 P.M.
b. 5:08 P.M.
c. 5:10 P.M.
d. 5:18 P.M.

Chapter 12

56. What is the approximate location of the Canary Islands hot spot?

a. 32° S 18° W
b. 32° S 18° E
c. 32° N 18° W
d. 32° N 18° E

57. Which process could lead directly to the formation of pumice rock?

a. precipitation of minerals from evaporating seawater
b. metamorphism of unmelted rock material
c. deposition of quartz sand
d. explosive eruption of lava from a volcano

58. Which best describes a major characteristic of both volcanoes and earthquakes?

a. They are centered at the Poles.
b. They are located in the same geographic areas.
c. They are related to the formation of glaciers.
d. They are restricted to the Southern Hemisphere.

59. Which is not a type of volcanic cone?

a. cinder
b. composite
c. shield
d. batholith

60. Which was formed by a hot spot?

a. the Himalayan mountains
b. the Hawaiian islands
c. the Andes mountains
d. the Mariana trench

Chapter 13

61. The Himalaya Mountains are located along a portion of the southern boundary of the Eurasian Plate. At the top of Mt. Everest (29,028 feet) in the Himalaya Mountains, climbers have found fossilized marine shells in the surface bedrock. From this observation, which statement is the best inference about the origin of the Himalaya Mountains?

- **a.** The Himalaya Mountains were formed by volcanic activity.
- **b.** Sea level has been lowered more than 29,000 feet since the shells were fossilized.
- **c.** The bedrock containing the fossil shells is part of an uplifted seafloor.
- **d.** The Himalaya Mountains formed at a divergent plate boundary.

62. Which is not a type of a fault?

- **a.** thrust
- **b.** reverse
- **c.** fold
- **d.** strike-slip

63. Which is the best evidence of crustal movement?

- **a.** molten rock in the Earth's outer core
- **b.** tilted sedimentary rock layers
- **c.** residual sediments on top of bedrock
- **d.** marine fossils found below sea level

64. Which features are commonly formed at the plate boundaries where continental crust converges with oceanic crust?

- **a.** large volcanic mountain ranges parallel to the coast at the center of the continents
- **b.** a deep ocean trench and a continental volcanic mountain range near the coast
- **c.** an underwater volcanic mountain range and rift valley on the ocean ridge near the coast
- **d.** long chains of mid-ocean volcanic islands perpendicular to the coast

65. Which of the following locations is the site of a convergent plate boundary?

- **a.** the mid-Atlantic ridge
- **b.** the Aleutian trench
- **c.** the Southeast Indian ridge
- **d.** the Pacific/North American plate boundary

Chapter 14

66. During which process does heat transfer occur because of density differences?

a. conduction
b. convection
c. radiation
d. reflection

67. Energy is transferred from the Sun to the Earth mainly by

a. molecular collisions.
b. density currents.
c. electromagnetic waves.
d. red shifts.

68. Which type of land surface would probably reflect the most incoming solar radiation?

a. light colored and smooth
b. light colored and rough
c. dark colored and smooth
d. dark colored and rough

69. Which is the major source of energy for most Earth processes?

a. radioactive decay within the Earth's interior
b. convection currents in the Earth's mantle
c. radiation received from the Sun
d. earthquakes along fault zones

70. All objects warmer than 0 K (absolute zero) must be

a. radiating electromagnetic energy.
b. condensing to form a gas.
c. warmer than 0° Celsius.
d. expanding in size.

Chapter 15

71. Summer days in New York State are likely to be hotter than winter days because in summer

a. Earth is closer to the Sun.
b. the number of sunspots increases.
c. Earth's northern axis is tilted toward the Sun.
d. the Sun gives off more energy.

72. How many times will the Sun's perpendicular rays cross Earth's equator between March 1 of one year and March 1 of the next year?

a. 1
b. 2
c. 3
d. 4

73. On June 21, some Earth locations have 24 hours of daylight. These locations are all between the latitudes of

a. 0° and 23.5° N
b. 23.5° N and 47° N
c. 47° N and 66.5° N
d. 66.5° N and 90° N

74. Electromagnetic radiation that reaches the Earth from the Sun is called

a. insolation.
b. conduction.
c. specific heat.
d. terrestrial radiation.

75. Which angle of the Sun above the horizon produces the greatest intensity of sunlight?

a. 70°
b. 60°
c. 40°
d. 25°

Chapter 16

76. Surface winds on Earth are primarily caused by differences in

a. air density due to unequal heating of Earth's surface.
b. ocean wave heights during the tidal cycle.
c. rotational speeds of Earth's surface at various latitudes.
d. distances from the Sun during the year.

77. The planetary winds in Earth's Northern Hemisphere generally curve to the right due to Earth's

a. orbit around the Sun.
b. spin on its axis.
c. magnetic field.
d. force of gravity.

78. A student used a sling psychrometer to measure the humidity in the air. If the relative humidity was 65% and the dry-bulb temperature was 10° C, what was the wet-bulb temperature?

a. 5° C
b. 7° C
c. 3° C
d. 10° C

79. An air mass originating over north central Canada would most likely be

a. warm and dry.
b. warm and moist.
c. cold and dry.
d. cold and moist.

80. Clouds usually form when moist air rises because the air

a. contracts and cools.
b. contracts and warms.
c. expands and cools.
d. expands and warms.

Chapter 17

81. Most of Earth's surface ocean current patterns are primarily caused by

a. the force of gravity.
b. the impact of precipitation.
c. river currents.
d. prevailing winds.

82. Which ocean current flows northeast along the eastern coast of North America?

a. Gulf Stream
b. North Equatorial
c. California
d. Labrador

83. Which two ocean currents are both warm currents that primarily flow away from the equator?

a. Guinea Current and Labrador Current
b. Brazil Current and Agulhas Current
c. Alaska Current and Falkland Current
d. Canaries Current and Gulf Stream Current

84. The flattest feature of the ocean basin is the

a. rift valley.
b. continental slope.
c. seamount.
d. sea floor.

85. Which isn't an abundant microscopic organism in the mixed layer?

a. phytoplankton
b. diatom
c. zooplankton
d. chemosynthetic bacteria

Chapter 18

86. Most water vapor enters the atmosphere by the processes of

a. convection and radiation.
b. condensation and precipitation.
c. evaporation and transpiration.
d. erosion and conduction.

87. Why are the beaches that are located on the southern shore of Long Island often considerably cooler than nearby inland locations on hot summer afternoons?

a. A land breeze develops due to lower specific heat of water and the higher specific heat of land.
b. A sea breeze develops due to the higher specific heat of water and the lower specific heat of land.
c. The beaches are closer to the Equator than the inland locations are.
d. The beaches are farther from the Equator than the inland locations are.

88. If the Earth's axis were tilted 35° instead of 23.5°, the average temperatures in New York State would most likely

a. decrease in both summer and winter.
b. decrease in summer and increase winter.
c. increase in summer and decrease in winter.
d. increase in both summer and winter.

89. Compared to an inland location, a location on an ocean shore at the same elevation and latitude is likely to have

a. cooler winters and warmer summers.
b. cooler winters and cooler summers.
c. warmer winters and warmer summers.
d. warmer winters and cooler summers.

90. Why do the windward sides of mountain ranges receive more precipitation than the leeward sides?

a. Rising air expands and cools.
b. Rising air compresses and cools.
c. Sinking air expands and cools.
d. Sinking air compresses and cools.

Chapter 19

91. The Milky Way galaxy is best described as

a. a type of solar system.
b. a constellation visible to everyone on Earth.
c. a region in space between the orbits of Mars and Jupiter.
d. a spiral-shaped formation composed of billions of stars.

92. Starlight from distant galaxies provides evidence that the universe is expanding because this starlight shows a shift in wavelength toward the

a. red-light end of the visible spectrum.
b. blue-light end of the visible spectrum.
c. ultraviolet-ray end of the electromagnetic spectrum.
d. gamma-ray end of the electromagnetic spectrum.

93. Which planet's orbit around the Sun is most nearly circular?

a. Mercury
b. Neptune
c. Pluto
d. Venus

94. Which list of three planets and Earth's Moon is arranged in order of increasing equatorial diameter?

- **a.** Earth's Moon, Pluto, Mars, Mercury
- **b.** Pluto, Earth's Moon, Mercury, Mars
- **c.** Mercury, Mars, Earth's Moon, Pluto
- **d.** Mars, Mercury, Pluto, Earth's Moon

95. Stars are classified on the H-R diagram by

- **a.** luminosity and temperature.
- **b.** mass and size.
- **c.** temperature and origin.
- **d.** luminosity and structure.

Chapter 20

96. Earth's orbital velocity is slowest on July 5 because

- **a.** the Moon is closest to Earth.
- **b.** Earth's distance from the Sun is greatest.
- **c.** Earth, the Moon, and the Sun are located along a straight line in space.
- **d.** the highest maximum temperatures occur in the Northern Hemisphere.

97. The apparent daily movement of the Sun across the sky is caused by

- **a.** Earth's rotation on its axis.
- **b.** Earth's revolution around the Sun.
- **c.** the Sun's revolution around the Earth.
- **d.** the Sun's rotation during a 24-hour period.

98. The length of an Earth year is based on Earth's

- **a.** rotation of 15° /hr.
- **b.** revolution of 15° /hr.
- **c.** rotation of approximately 1° /day.
- **d.** revolution of approximately 1° /day.

99. The motion of a Foucault pendulum provides evidence of

- **a.** the Sun's rotation.
- **b.** the Sun's revolution.
- **c.** Earth's rotation.
- **d.** Earth's revolution.

100. The passage of the Moon into Earth's shadow causes a

- **a.** lunar eclipse.
- **b.** solar eclipse.
- **c.** new Moon.
- **d.** full Moon.

Chapter 21

101. Why are radioactive substances useful for measuring geologic time?

- **a.** The ratio of decay products to radioactive substances remains constant in rocks.
- **b.** The half-lives of radioactive substances are short.
- **c.** Samples of radioactive substances are easy to collect from rocks.
- **d.** Radioactive substances undergo decay at a predictable rate.

102. Which radioactive isotope is most useful for determining the age of mastodont bones found in late Pleistocene sediments?

- **a.** uranium-238
- **b.** carbon-14
- **c.** potassium-40
- **d.** rubidium-87

103. Older layers of rock may be found on top of younger layers of rock as a result of

- **a.** weathering processes
- **b.** igneous extrusions
- **c.** joints in the rock layers
- **d.** overturning of rock layers

104. A sample of wood found in an ancient tomb contains 25% of its original carbon-14. The age of this wood sample is approximately

- **a.** 2,800 years.
- **b.** 5,700 years.
- **c.** 11,400 years.
- **d.** 17,100 years.

105. What is the relative age of a fault that cuts across many rock layers?

- **a.** The fault is younger than all the layers it cuts across.
- **b.** The fault is older than all the layers it cuts across.
- **c.** The fault is the same age as the top layer it cuts across.
- **d.** The fault is the same age as the bottom layer it cuts across.

Chapter 22

106. Approximately what percentage of the estimated age of Earth does the Cenozoic Era represent?

a. 1.4%
b. 5.0%
c. 11.9%
d. 65.0%

107. Which group of organisms, some of which were preserved as fossils in early Paleozoic rocks, are still in existence today?

a. brachiopods
b. eurypterids
c. graptolites
d. trilobites

108. There is evidence that an asteroid or a comet crashed into the Gulf of Mexico at the end of the Mesozoic Era. Consequences of this impact event may explain the

a. extinction of many kinds of marine animals, including trilobites.
b. extinction of ammonoids and dinosaurs.
c. appearance of the earliest birds and mammals.
d. appearance of great coal-forests and insects.

109. Which type of rock most likely contains fossils?

a. scoria
b. gabbro
c. schist
d. shale

110. In order for an organism to be used as an index fossil, the organism must have been geographically widespread and must have

a. lived on land.
b. lived in shallow water.
c. existed for a geologically short time.
d. been preserved by volcanic ash.

Answers: **1.**a, **2.**a, **3.**c, **4.**c, **5.**b, **6.**b, **7.**a, **8.**b, **9.**a, **10.**a, **11.**c, **12.**b, **13.**c, **14.**a, **15.**d, **16.**d, **17.**a, **18.**c, **19.**a, **20.**d, **21.**d, **22.**a, **23.**b, **24.**d, **25.**b, **26.**d, **27.**b, **28.**d, **29.**b, **30.**a, **31.**b, **32.**c, **33.**d, **34.**c, **35.**c, **36.**a, **37.**b, **38.**d, **39.**d, **40.**a, **41.**d, **42.**a, **43.**d, **44.**c, **45.**d, **46.**c, **47.**c, **48.**a, **49.**d, **50.**b, **51.**d, **52.**a, **53.**c, **54.**a, **55.**a, **56.**c, **57.**d, **58.**b, **59.**d, **60.**b, **61.**c, **62.**c, **63.**b, **64.**b, **65.**b, **66.**b, **67.**c, **68.**a, **69.**c, **70.**a, **71.**c, **72.**b, **73.**d, **74.**a, **75.**a, **76.**a, **77.**b, **78.**b, **79.**c, **80.**c, **81.**d, **82.**a, **83.**b, **84.**d, **85.**d, **86.**c, **87.**b, **88.**c, **89.**d, **90.**a, **91.**d, **92.**a, **93.**d, **94.**b, **95.**a, **96.**b, **97.**a, **98.**d, **99.**c, **100.**a, **101.**d, **102.**b, **103.**d, **104.**c, **105.**a, **106.**a, **107.**a, **108.**b, **109.**d, **110.**c

CQR RESOURCE CENTER

CQR Resource Center offers the best resources available in print and online to help you study and review the core concepts of Earth Science.

Books

Earth Science, 1st edition, by Christie Borgford *et al* (Holt, Rinehart, and Winston, 2001) covers the wide range of Earth Science topics in a clear manner with easy to understand text and great diagrams.

Earth Science, 1st editon, by Ralph Feather, Jr. and Susan Heath Snyder (Glencoe/McGraw-Hill, 1997) correlates to the National Science Content Standards. The descriptions and diagrams are well laid out and easy to understand.

Earth Science, 4th edition, by Samuel Namowitz and Nancy Spaulding (D.C. Heath and Company, 1989) has higher level text than the other books mentioned but covers the material equally well.

Internet

Visit the following Web sites for more information about the Earth Science topics listed.

Astronomy

Sky and Telescope—http://skyandtelescope.com/—contains up-to-date information covering a wide range of topics.

United States Naval Observatory—http://aa.usno.navy.mil/data/—has lots of data such as moon phases, eclipses, and transits.

Space.com—http://www.space.com/—has daily lists of current articles about a variety of top stories related to astronomy.

Hubble Space Telescope—http://www.stsci.edu/hst/—has the latest pictures and information about the Hubble telescope.

Geology

USGS Earthquake hazards program—http://earthquake.usgs.gov/learning/kids.php?—has information about earthquakes for kids.

National Snow and Ice Data Center—http://nsidc.org/glaciers/—has extensive information about glaciers.

Improbable Research—http://www.improbable.com/airchives/paperair/volume9/v9i3/kansas.html—has an article titled "Kansas is flatter than a pancake" and does research to support the title of the article.

Terraserver.com—http://www.terraserver.com/—contains satellite images of the Earth.

United States Geological Survey—homepage http://www.usgs.gov/—contains lots of information and links related to the field of geology.

Meteorology

USA Today Weather—http://asp.usatoday.com/weather/weather front.aspx—has the lastest weather forecasts.

Web Weather for Kids—http://eo.ucar.edu/webweather/—has lots of information and activities for kids.

Hurricane tracking for the Atlantic and Pacific Ocean—http://hurricane.terrapin.com/—has the latest information for tropical storms.

American Meteorological Society—http://www.ametsoc.org/—is the homepage for this organization.

BBC Weather—http://www.bbc.co.uk/weather/world/—provides weather information for the world.

NOAA/ National Climatic Data Center—http://www.ncdc.noaa.gov/oa/ncdc.html—contains extensive climate information.

National Weather Association—http://www.nwas.org/—is the homepage for this organization.

Isaac's Storm—http://www.randomhouse.com/features/isaacsstorm/—contains information about the 1900 Galveston hurricane known as Isaac's storm.

Royal Meteorological Society—http://www.royal-met-soc.org.uk/cweb.html—is the homepage for this society.

Tropical storm tracks: yearly summaries—http://www.solar.ifa.hawaii.edu/Tropical/summary.html—contains maps of storm tracks listed by year.

The Weather Channel—http://www.weather.com—is the homepage for this broadcasting company.

NOAA/National Weather Service—http://www.weather.gov/—is the homepage for the government's weather branch.

World Meteorological Organization—http://www.worldweather.org/—is the homepage for the organization responsible for official forecasts worldwide.

Assorted Science Sites

Discovery Channel—http://dsc.discovery.com/—covers a wide range of topics.

NASA-Earth Observatory—http://earthobservatory.nasa.gov/—has new satellite imagery and scientific information about Earth.

American Museum of Natural History—http://ology.amnh.org/—is the homepage for the Museum in New York City.

ScienceWeek—http://scienceweek.com/—this online publication covers recent topics and discoveries.

Discover magazine—http://www.discover.com/—is the homepage for their online magazine.

US Environmental Protection Agency—http://www.epa.gov/—covers topics concerning human health and the environment.

FEMA Web site for kids—http://www.fema.gov/kids/dizarea.htm—contains a section called "The disaster area," covering information about many natural disasters.

NY State Regents—http://www.nysedregents.org/—past Regents exams are posted here.

Oswego City School District Regents Exam Prep Center—http://www.regentsprep.org/—has review questions for Earth Science (and many other subjects) by topic.

Scientific American magazine—http://www.scientificamerican.com/—is the homepage for this online magazine.

GLOSSARY

The numbers in parenthesis after each entry in the glossary refer to the chapter where you will find the term discussed.

abrasion (6) the process of particles rubbing against each other, wearing down the sharp edges into smaller pieces.

absolute magnitude (19) how bright a star appears from 32.6 light years away.

absolute zero (14) temperature at which all molecular motion ceases (0°K, –273.15°C, or –459.67°F).

acid test (4) tests a mineral or rock for calcite or calcium in its composition.

active continental margin (13) plate boundary where mountains are formed.

active volcano (12) volcano that has erupted in recorded history.

advected fog (16) fog caused by air moving from one region to another region with a different surface temperature.

advection (14) movement of air from one area to another, creating unusually warm or cold conditions.

air mass (16) large mass of air at the surface of the Earth with similar characteristics of temperature and humidity throughout the mass.

albedo (18) reflection of sunlight.

altitude (20) angle of a celestial object above the horizon.

anemometer (16) instrument used to determine wind speed.

anticline (13) parallel rock layers folded upward, like an arch.

aphelion (20) point in an orbit farthest from the Sun.

apogee (20) point in an orbit farthest from the Earth.

apparent diameter (20) the diameter of an object as it appears to an observer; changes with distance.

apparent magnitude (19) how bright a star appears to someone on Earth.

arid (8) climates that are extremely dry, with low precipitation and high evapotranspiration.

asteroid belt (19) the orbiting band of rocks between Mars and Jupiter.

asthenosphere (10) the outer layer of the Earth's mantle, which has a plastic-like composition; site of convection currents that move the plates on the surface.

astronomical unit (19) measurement used in the solar system, the average distance from the Earth to the Sun (150,000,000 km or 93,000,000 mi).

atmosphere (2) the gas layer surrounding the Earth.

atoll (17) ring of coral reefs surrounding a sunken island.

aurora (19) light created in the sky by the interference of charged solar particles with the magnetic field of the Earth.

autumnal equinox (15) September 23, when the vertical ray of the Sun is at the Equator; the entire Earth has 12 hours of day and 12 hours of night.

azimuth (20) direction along the horizon when looking for an object in the sky.

big-bang hypothesis (19) says that origin of the universe was when all matter was collected together and exploded, about 15 billion years ago.

blue shift (19) apparent shortening of starlight as it moves toward you (Doppler).

buoyant (16) describes a property that causes materials to appear to float in or on a fluid.

caldera (12) extinct volcano that has a collapsed cone.

capillary (9) water that works its way upward in the ground through adhesion to rock particles and cohesion to itself.

carbonates (4) the family of minerals composed of carbon and oxygen.

Cephid (19) a star that varies in its light output.

chemical weathering (6) the breaking down of rock material by chemical means, forming a new substance.

chromosphere (19) reddish layer of the Sun that is hydrogen burning.

cinder cone (12) steep cone of a volcano formed from ash and loose rock.

cirrus (16) very high clouds formed by ice crystals; look like feathers; usually associated with fair weather.

clastic (5) pieces of rocks.

cleavage (4) the way a mineral splits or breaks along weak bonds in planes.

climate (18) the overall temperature, precipitation, and weather conditions for an area.

cold front (16) the leading edge of a cold air mass.

color (4) an easy test in the identification of minerals, but not always reliable.

column (9) pillar of rock formed when a stalactite and stalagmite merge.

comet (19) a mass of frozen gases, ice, and rock that orbits the Sun.

composite volcano (12) a volcano consisting of a cone of alternating layers of solidified lava and rock particles.

compound (4) a molecule made up of two or more elements.

condensation (14) the process of a substance changing states from gas to liquid.

conduction (14) transfer of energy through solids by direct contact.

constellation (19) a group of stars that appears to be in a pattern.

contact metamorphism (5) the process of changing rocks into metamorphic rocks by nearby magma.

continental glacier (6) a thick sheet of ice covering a mass of land all year round, moving outward from the thickest part; found only in Greenland and Antarctica.

continental rise (17) area which connects the continental slope to the ocean floor.

continental shelf (17) ocean bottom along the coasts with a low gradient.

continental slope (17) ocean bottom that connects the continental shelf to the deep ocean floor.

contour farming (6) planting crops along contour lines on a hill to help prevent soil erosion.

contour interval (3) the elevation difference between contour lines.

contour line (3) line connecting points of equal elevation.

convection (14) transfer of energy in fluids; can create currents by density differences.

converging plate boundary (10) region where plates move toward each other.

coordinate system (1) rectangular grid system for plotting points.

Copernicus, Nicholas (20) astronomer who developed the heliocentric model.

corona (19) outermost layer surrounding the Sun.

crater (19) impact mark left on a planet or moon by a collision with a another object.

craton (13) core rock base of a continent.

creep (6) slow movement of rocks and sediments down a hillside.

crust (10) the thin, outer layer of the solid part of the Earth.

crystal shape (4) the arrangement of molecules in a mineral.

cumulonimbus (16) cumulus cloud that builds vertically; usually associated with a cold front and thunderstorms.

cumulus (16) puffy, cotton-like clouds formed by rising air.

cyclogenesis (16) the process of forming storm systems.

deficit (9) time during the water budget when drought conditions occur; there is not enough water to supply the needs of an area.

delta (7) a triangle-shaped area of deposition located at the mouth of an older river.

density (1) the amount of matter in a given space.

deposition (14) the rare process of a substance changing states from gas directly to a solid.

derived unit (1) unit that is a combination of basic units.

dew point (16) temperature at which water vapor condenses into liquid water.

discharge (6) the amount of water in a river or stream that passes a certain point in a given amount of time.

diverging plate boundary (10) region where plates move apart.

doldrums (16) region along the Equator where the weather conditions are fairly consistent.

dormant volcano (12) volcano that has not erupted during recorded history.

drainage basin (8) the area around a stream that could drain into the stream.

drainage divide (8) the outer edge between drainage basins.

drift (7) small particles carried away from larger rocks by glacial meltwater.

drizzle (16) liquid precipitation that is less than .02 cm in diameter.

dry adiabatic lapse rate (16) rate at which dry air cools as it rises upward.

dry-bulb temperature (16) temperature of the air.

dynamic equilibrium (1) a system that fluctuates, but overall is in balance.

El Niño/La Niña (18) long-term weather patterns associated with changing global winds and ocean temperatures in the Pacific Ocean.

electromagnetic spectrum (19) diagram that breaks down energy by wavelength.

epicenter (11) location on the Earth's surface directly above the focus of an earthquake.

epicycle (20) small circle made by a planet's orbit in the geocentric model.

equilibrium (1) a system that is in balance.

erosion (6) the transportation of weathered materials.

erratic (7) a large rock that is deposited by a glacier.

explosive eruption (12) volcanic activity containing thick lava and more gases under pressure; creates steeper cones.

extrusive (5) rock that forms on the Earth's surface.

eye (16) center of circulation in a hurricane, where the conditions are calm.

eye wall (16) area surrounding the eye of a hurricane, where winds are most intense.

fault (13) a crack or break in a rock.

fault plane (13) the surface that rocks move along when plates shift.

fault-block mountains (13) mountain range formed when sections of sedimentary rocks are tilted upward in sections.

felsic (5) magma that is aluminum or silica based, lighter in color, and less dense than mafic magma.

fetch (6) length of open ocean for wind to create waves.

field (3) an area where there is a measurable amount of a specific value at every point.

flood plain (7) the area along the banks of a meandering river that are prone to flooding at various times.

focus (11) exact site of the origin of an earthquake, below the epicenter.

folded mountains (13) mountain range formed by the collision of continental plates, causing the rock layers to be crumpled.

foliated texture (5) layered mineral crystals in a metamorphic rock.

fossil (22) evidence that life was present, preserved in a rock.

fracture (4) the uneven breaking or cracking of a mineral.

front (16) the boundary between two air masses.

frost point (16) dew point temperature, below 0°C.

Fujita scale (16) scale for measuring the intensity of a tornado from F0 to F5.

Galileo (20) scientist whose observations supported the Copernican model.

geocentric (20) a model in which Earth is at the center of the universe.

geyser (12) groundwater deep in the Earth that turns to steam and is forced to the surface when enough pressure is generated.

granule (19) one of the individual cells that make up the photosphere of the sun.

graphical model (2) a graph showing relationships.

greenhouse effect (19) incoming (shortwave) energy is reradiated as heat energy (longwave) and trapped by the greenhouse gases in the atmosphere; causes global warming.

greenhouse gas (1) a gas produced from burn fossil fuels, which hold and trap heat energy; carbon dioxide is an example.

ground fog (16) fog formed by radiational cooling.

guyot (17) flat-topped seamount, eroded by ocean waves while exposed.

hail (16) layers of ice formed in a large thunderstorm cloud that fall to Earth.

half-life (21) the amount of time it takes a radioactive element to decay into half of its original mass.

heliocentric (20) sun-centered model of the solar system.

Horse Latitudes (16) area of little surface winds and high pressure along 30° N latitude.

hot spot (10, 12) weak or thin area in a plate that allows magma to rise up and reach the surface.

hot spring (12) groundwater heated by magma rising to the surface through an opening in the ground.

hurricane (16) low-pressure system with sustained winds of 74 mph or greater.

hydrosphere (2) the water layer of the Earth.

hygrometer (16) an instrument used to measure humidity; can be made from hair.

igneous (5) rock formed when molten rock cools.

impermeable (9) rock that water cannot sink into or through.

index fossil (21) a fossil of an organism that occurred for a short amount of time and over a widespread area; used to connect rock layers over long distances.

inference (1) an educated guess based on collected data.

infiltration (9) water sinking into the ground.

inner core (10) the solid center of the Earth, composed of iron and nickel.

insolation (15) a combination of the words "incoming solar radiation."

instrument (1) tool used to extend your senses and gather data.

International Date Line (3, 20) the longitude line measuring 180° where the day changes.

Intertropical Convergence Zone (16) area along the Equator where trade winds from the Northern and Southern hemispheres meet, forming thunderstorms.

intrusion (12) magma that crosses through other rock layers, cooling and hardening before reaching the surface.

intrusive (5) rock that forms below the Earth's surface (plutonic).

inversion (18) an atmospheric condition where warm air is on top of cold air.

isobar (16) line connecting areas of equal pressure on a map.

isoline (3) line that connects equal values.

isosurface (3) three-dimensional diagram showing surfaces connecting equal values.

isotherm (16) line connecting areas of equal temperature on a map.

isotopes (21) two elements that have the same atomic number but different atomic masses.

jet stream (16) band of fast-moving air in the upper troposphere.

Jovian planet (19) outer planet (Jupiter, Saturn, Uranus, and Neptune) consisting of a gaseous surface.

Kepler, Johnannes (20) astronomer who developed three laws of planetary motion: planets move in elliptical orbits; planets sweep out equal area in equal time; the period of revolution is proportional to the distance to the Sun.

kettle (7) a steep-sided pond created by a glacier when a block of ice is left behind as the glacier retreats.

key bed (21) a layer in the rock record showing an event that occurred quickly and over a widespread area; is used like an index fossil.

land breeze (16) local wind that forms at night along a beach due to uneven cooling rates of land and water, wind moves from land to water.

lateral moraine (7) the pile of rocks that accumulate along the edges of a glacier.

latitude (3) coordinate lines for locating a position on Earth that run east and west and are parallel to each other, running from 0° to 90° (angle from the equator).

lava (5) liquid rock on the surface of the Earth.

leeward (18) the side opposite from the prevailing wind direction.

lifting condensation level (16) formula used to find the height at which clouds can form.

light year (19) the distance that light travels in one year, about 9.5 trillion kilometers.

lightning (16) electricity generated by a thunderstorm.

line graph (1) points plotted on a coordinate system and connected with a line.

lithosphere (2, 10) the rock layer on the outer edge of the Earth.

local noon (3) occurs when the Sun is at its highest point for the day.

longitude (3) coordinate lines for locating position on Earth that run north and south through the poles, are farthest apart at the Equator, run from 0° to 180°, are equal in length, and are measured from the Prime Meridian.

longshore current (6) a current moving parallel to the coast.

low-pressure center (16) counterclockwise circulation center, formed along a stationary front in the Northern Hemisphere.

luminosity (19) actual brightness of a star.

lunar eclipse (20) the Moon goes out of view as it moves into the Earth's shadow; occurs during the Full Moon phase.

luster (4) the way a mineral shines in reflected light.

L-wave (11) longitudinal wave created by the P-wave and S-waves of an earthquake at the surface of the Earth; these are the slowest and move outward like ripples on a pond.

mafic (5) magma that is iron or magnesium-based, darker in color, and more dense than felsic magma.

magma (5) liquid rock below the Earth's surface.

magnetic declination (3) the number of degrees that a compass needle is pulled away from True North to point toward Magnetic North.

Magnetic North (3) the area near Hudson Bay, Canada, where the Earth's magnetic field is strongest in the Northern Hemisphere (location changes over time).

main sequence star (19) star that falls into broad band along the H-R diagram.

mantle (10) the layer below the crust; about 2,900 km thick; contains rocks rich in iron, magnesium, and silicon.

maria (19) large, flat areas on the Moon.

Marianas Trench (17) deepest trench in the world; lies off the coast of Japan.

mathematical model (2) an equation representing an idea.

matter (4) anything that has mass and takes up space.

mechanical model (2) a physical model with moving parts.

mechanical weathering (6) the physical breaking down of rock, changing only its size (smaller); examples are ice wedging, plant action, and pressure unloading.

medial moraine (7) the moraine created when two glaciers meet and their lateral moraines merge.

mental model (2) an idea or model that exists in your mind.

Mercalli Scale (11) scale for measuring earthquakes based on obsevations.

meridian (3) line of longitude.

mesopause (16) region between the mesosphere and the thermosphere.

mesosphere (16) part of the atmosphere between the stratosphere and the thermosphere; temperatures decrease with altitude.

metamorphic (5) existing rock that undergoes extreme heat or pressure and is recrystallized.

meteor (19) a meteoroid that enters the Earth's atmosphere; also known as a shooting star.

meteorite (19) a meteor that reaches the surface of the Earth.

meteoroid (19) a rock fragment orbiting in the solar system.

mid-latitude low (16) low-pressure system that forms along a stationary front.

mid-ocean ridge (10) area between two diverging plates, where magma reaches the surface and causes an area of increased elevation and new crust along the ocean floor.

Milankovic period (18) long term climate changes due the wobble of the Earth's revolution around the Sun.

Milky Way galaxy (19) the spiral galaxy that includes our solar system near its outer edge.

model (2) a scale representation of another object or idea.

Moh's hardness scale (4) a relative scale ranging from 1 to 10, measuring whether a mineral can scratch another mineral.

moist adiabatic lapse rate (16) rate at which humid air cools as it rises upward.

monsoon (16) seasonal wind pattern changes that cause rainy and dry seasons.

moraine (7) large area of deposition left behind from the advance of a glacier.

natural selection (22) theory by Charles Darwin which states that organisms best fit for their environment will survive and pass along these traits.

neap tide (20) tide with the smallest tidal range; during both quarter phases.

nebula (19) a cloud of gas and dust in space.

Newton, Issac (20) developed the universal law of gravity.

nonrenewable resources (1, 14) natural resources that cannot be replenished for millions of years, if at all.

normal fault (13) fault that occurs when rocks are pulled apart, causing one side to move downward.

oblate spheroid (2) the shape of the Earth; not a perfect sphere, but flatter at the poles and slightly bulging at the Equator.

observation (1) data that is collected through your senses.

occluded front (16) front that forms when a cold front catches up to a warm front.

orogeny (22) a mountain-building period or event.

outer core (10) the iron- and nickel-rich liquid layer near the center of the Earth.

outgassing (22) gases that escape to the atmosphere during volcanic eruptions.

outwash plain (7) sandy area downstream from a moraine created by drift particles and meltwater from a glacier.

overriding plate (10) crustal plate which collides with a more dense plate and moves above the plate.

overturning (13) rock layers that are flipped upside-down during the mountain-building process.

parallax (20) the apparent change in the position of stars due to Earth's revolution.

parallels (3) lines that never meet, such as latitude.

passive continental margin (13) area along the coast where sediments are deposited.

pause (2) a region separating two layers of the atmosphere.

penumbra (20) the lighter, gray area of a shadow.

percent error (1) formula used to find the inaccuracy of a measurement.

perigee (20) position in an orbit that is closest to the Earth.

perihelion (20) point in an orbit that is closest to the Sun.

permeability (9) the rate at which water sinks into the ground.

photosphere (19) outer layer of the Sun.

physical model (2) a representation of an object that you can hold.

phytoplankton (17) microscopic plants floating freely in the ocean.

plutonic (5) rock that forms below the Earth's surface (intrusive).

Polaris (3) the North Star, located in the Little Dipper constellation.

pollution (1) a substance that harms living organisms or the environment.

porosity (9) the percentage of open space between soil particles/rocks.

porphyritic texture (5) rocks that have different-sized crystals, created at different times.

prevailing westerlies (16) planetary winds between 30° and 60° latitude; they guide weather systems for the United States (from SW to NE).

prime meridian (3) 0° longitude line, running though Greenwich, England.

profile (3) side view of land on a topographic map.

psychrometer (16) an instrument with a dry-bulb thermometer and a wet-bulb thermometer; used to measure dewpoint and relative humidity.

Ptolemy (20) astronomer who developed the geocentric model.

pulsar (19) a star that sends energy out in pulses.

punctuated equilibrium (22) theory that organisms just "appear" on Earth.

P-waves (11) primary waves generated by an earthquake; these compression waves are the fastest and travel through solids, liquids, and gases.

radiation (14) transfer of energy through a vacuum; the way in which the Sun supplies the Earth with energy.

radioactive element (21) an atom that emits gamma rays, alpha particles, and beta particles; can be used to determine age.

radiosonde (16) weather balloon that carries instruments to record data and transmit readings to a base unit.

rain (16) liquid precipitation larger than .02 cm in diameter.

recharge (9) the period of time during the water-budget cycle when the water in the ground is being replenished.

red giant (19) stage of the life cycle of a star in which it expands and cools.

red shift (19) apparent lengthening of starlight as it moves away from you (Doppler).

regionally metamorphic (5) rock that undergoes intense heat and pressure over large areas.

relative humidity (16) a measure of how much water is in the air compared to how much it can hold, given as a percentage.

renewable resources (1, 14) energy sources or other natural resources that are replenished shortly after being used.

retrograde motion (19) the apparent backward movement of a celestial object.

reverse fault (13) rocks that crack and are thrust upward, forming an overhang; caused by the compression of rocks.

Richter Scale (11) scale for measuring earthquakes based on energy released.

rift eruption (12) lava flows in long, narrow cracks of the Earth's crust.

rift valley (10) the space between diverging plates.

rip currents (6) strong surface currents that move like small rivers perpendicular to the shoreline, caused by water returning from the beach to the ocean.

rock cycle (5) the continuous flow from one type of rock to another.

Saffir-Simpson scale (16) scale for measuring hurricane intensity, from Category I to Category V.

salinity (17) amount of dissolved salts in water.

saturated (16) cannot hold any more water, as in groundwater or air (humidity).

scalar (3) field values that measure magnitude.

scale (2) the proportion of a model in relation to the original.

scientific notation (1) a method of converting very large or very small numbers into a convenient value using exponents.

sea breeze (16) local wind that develops during the day along a beach due to uneven heating of land and water; wind moves from water to land.

seamount (17) underwater mountain.

sedimentary (5) rock formed from the compaction and cementation of fragments from other rocks.

seismogram (11) paper record graphing earthquake motions, created by a seismograph.

seismograph (11) machine that detects earthquakes.

severe thunderstorm (16) thunderstorm that has winds in excess of 50 mph and can produce large hail.

shadow zone (11) area of the Earth shielded from earthquake waves by the outer core (where S-waves are absorbed and P-waves are refracted).

shield cone (12) broad cone of a volcano resulting from smooth lava flows.

silicates (4) group of minerals with silicon and oxygen as a base.

sliding plate boundary (10) region where plates move next to each other.

sling psychrometer (16) a handheld psychrometer than spins around, used to measure dewpoint and relative humidity.

slip face (7) the back side of a sand dune.

slump (6) rock material that is moved downhill as a block of land is uplifted.

solar eclipse (20) event in which the view of the Sun is blocked by the Moon during a New Moon phase, when the Moon's shadow reaches the Earth.

solar noon (20) the highest point of the Sun on any day.

solar prominence (19) flame-like arc extending out from the Sun.

specific gravity (4) the relative density of a mineral, compared to water.

specific humidity (16) a measure of how much water is actually in the air.

spectroscope (19) instrument for separating visible light into colors.

spiral galaxy (19) galaxy with bands of stars that rotate around the center.

spring tide (20) extreme tides during Full and New Moon phases.

stalactite (9) rock icicle forming from the ceiling of a cave.

stalagmite (9) rock formation rising up from a cave floor.

station model (16) a diagram on a weather map showing weather data from a specific city at a given time.

stationary front (16) initial boundary between two air masses.

storm surge (16) bubble of water carried by a hurricane, causing coastal flooding.

stratopause (16) the region between the stratosphere and the mesosphere.

stratosphere (16) the layer of the atmosphere above the troposphere; temperature increases with altitude.

stratus (16) layered, sheet-like clouds, usually associated with warm fronts and found at lower altitudes.

streak (4) the color of the powder left behind when a mineral is rubbed along an unglazed porcelain tile.

striations (6, 7) parallel scratches on the Earth's surface caused by rocks dragged by a glacier; the scratches point in the direction of the glacial movement.

strike-slip fault (13) a fault where rock plates move horizontally to each other.

subducting plate (10) crustal plate which collides with another plate and moves under it.

sublimation (14) the rare process of a substance changing states from solid directly to a gas.

subsidence (17) land sinks into the sea; opposite of uplifting.

summer solstice (15) June 21, when the vertical ray of the Sun is at the Tropic of Cancer (23.5° N latitude), and is the longest day of the year in the Northern Hemisphere.

sunspot (19) cooler area on the surface of the Sun.

supercell (16) thunderstorm with strong updrafts that exist for hours and can spawn tornadoes.

supernova (19) the explosion of a star.

surplus (9) the period of time when the precipitation exceeds the needs for an area and the ground is saturated; runoff occurs, causing flooding conditions.

S-waves (11) secondary waves created by an earthquake; these shearing waves move at right angles to the path of travel and are stronger than P-waves, but only move through solids.

syncline (13) parallel rock layers folded downward in a valley-like formation.

talus (6) a pile of rocks at the bottom of a hill.

terminal moraine (7) a moraine created at the end of the advance of a glacier.

terrane (13) large pieces of rock that are moved large distances; can be from another plate.

terrestrial planet (19) inner planet (Mercury, Venus, Earth, and Mars) with a rocky surface.

thermocline (17) zone in water where the temperature changes drastically.

thermosphere (16) uppermost layer of the atmosphere; temperatures increase with altitude.

thin-skinned thrusting (13) thin, horizontal sheets of rock from the edge of a continent are forced inland.

till (7) an unsorted pile of sediment left behind when a glacier melts.

topographic map (3) map that shows elevations above sea level.

tornado (16) narrow, funnel-shaped column of wind created by a thunderstorm.

trace fossil (22) evidence that life existed in the past, such as footprints and burrows.

trade winds (16) planetary winds between 0° and 30° latitude.

transform plate boundary (10) region where plates move next to each other.

transparent (18) light can pass through the substance.

trench (10, 17) a deep canyon in the ocean caused by a plate being subducted under another plate.

tropical depression (16) strong low-pressure system formed in low latitudes.

tropical storm (16) low-pressure system with sustained winds from 39 to 73 mph.

tropopause (16) region between the troposphere and the stratosphere.

troposphere (16) lowest level of the atmosphere, where weather occurs; temperature decreases with altitude.

True North (3) geographic North Pole, latitude 90° N.

tsunami (6) a large wave created by an underwater earthquake or landslide.

turbidity currents (17) undersea mudslides.

umbra (20) the darkest part of a shadow.

unconformity (21) a break or gap in the rock record where layers of rock have been eroded.

universal law of gravity (20) formula that measures the force of gravity; developed by Isaac Newton.

updraft (16) wind current moving upward in a cloud.

upwelling (17) deep ocean water is pulled to the surface by currents.

urban heat island effect (18) city areas are warmer than suburbs or rural areas due to less vegetation, more land coverage and other infrastructure.

usage (9) time in the water budget for an area when water is being used faster than it is being replenished.

valley glacier (6) a glacier located on top of a mountain, also called an alpine glacier, that moves downhill through the valley.

vaporization (14) the process of boiling.

varve (21) alternating layers of sediment showing yearly cycles.

vector (3) field quantity that measures magnitude and direction.

Vernal equinox (15) March 21, when the vertical ray of the Sun is at the Equator; the entire Earth has 12 hours of day and 12 hours of night.

vertical ray (16) radiational energy from the Sun that strikes the Earth at a right angle.

vesicular texture (5) rock that has gas pockets and air that was trapped during the rock's formation.

waning (20) part of the lunar cycle in which the Moon is getting less full and the "left side is lit up" (left side lit).

warm front (16) the leading edge of a warm air mass.

water table (9) the surface of the water in the zone of saturation.

waxing (20) part of the lunar cycle when the Moon is getting more full, and the "right side is lit up."

weathering (6) the physical or chemical breaking down of rocks due to exposure to the atmosphere.

wet-bulb depression (16) difference between wet-bulb and dry-bulb temperatures, used to find relative humidity and dew point temperatures.

white dwarf (19) final stage of a star as it collapses onto itself.

windward (18) the side facing the wind.

Winter solstice (15) December 21, when the vertical ray of the Sun is at the Tropic of Capricorn (23.5° S latitude), the shortest day of the year in the Northern Hemisphere.

zenith (20) the point directly overhead of an observer.

zone of accumulation (6) upper level of a glacier where more snow falls than melts.

zone of aeration (9) area underground above the water table where the spaces between rocks contain a mixture of air and water.

zone of saturation (9) area underground where the spaces between rock particles are filled with water.

Index

D

E

F

G

H

I

J

K

L

M

Q

R

S

T

U

V

W

X

Y

Z

Notes